The Mind Made Flesh

Nicholas Humphrey, School Professor at the London School of Economics and Professor of Psychology at the New School for Social Research, New York, is a theoretical psychologist, internationally known for his work on the evolution of human intelligence and consciousness. His books include *Consciousness Regained*, *The Inner Eye*, *A History of the Mind*, and *Leaps of Faith*. He has been the recipient of several honours, including the Martin Luther King Memorial Prize and the British Psychological Society's book award.

OXFORD
UNIVERSITY PRESS

Great Clarendon Street, Oxford OX2 6DP

Oxford University Press is a department of the University of Oxford
It furthers the University's objective of excellence in research, scholarship,
and education by publishing worldwide in

Oxford New York

Athens Auckland Bangkok Bogotá Buenos Aires
Cape Town Chennai Dar es Salaam Delhi Florence Hong Kong Istanbul
Karachi Kolkata Kuala Lumpur Madrid Melbourne Mexico City Mumbai
Nairobi Paris São Paulo Shanghai Singapore Taipei Tokyo Toronto Warsaw

with associated companies in Berlin Ibadan

Oxford is a registered trade mark of Oxford University Press
in the UK and in certain other countries

Published in the United States
by Oxford University Press Inc., New York

British Library Cataloguing in Publication Data
Data available

Library of Congress Cataloging in Publication Data
Data available

ISBN 0–19–280227–5 (pbk.)

1 3 5 7 9 10 8 6 4 2

Typeset by Invisible Ink
Printed in Great Britain
on acid-free paper by
Biddles Ltd.,
Guildford and King's Lynn

For Ada and Samuel

Preface

This is a volume of essays, lectures, journal entries, newspaper articles that I have written in response to opportunity, when something happened that set me thinking on new lines—a surprise turn in my life, a serendipitous discovery, a left-field thought, a provocation or an invitation that could not be refused. The themes are various and the lines do not all converge. Still, these chapters do have this in common: they all concern the uneasy relation between minds and bodies, and they all take issue with received ideas.

In 1983 Oxford University Press published a previous volume of my essays, under the title *Consciousness Regained: Chapters in the Development of Mind*. I wrote in the Preface: 'The theme that runs through this collection is a concern with *why* human beings are as they are . . . The answers that most interest me are historical and evolutionary. Human beings are as they are because their history has been (so we may guess) as it has been.'

Twenty years—and four books—later, my interests are still centred in evolutionary psychology: a field which had yet to be named, but which has emerged as the most fertile field of all psychology. I have continued to be fascinated by the perennial issues: consciousness, justice, social understanding, spirituality. I hope I have better answers to some of the philosophical issues that puzzled me then. Some political and social problems have disappeared from view, others remain as pressing as ever, new ones have emerged.

My grandfather, A. V. Hill, in the Preface to his own collected writings (*The Ethical Dilemma of Science*, 1960), quoted Samuel Johnson: 'Read over your compositions and where ever you meet with a passage which you think is particularly fine, strike it out.' I cannot claim to have followed this advice to the letter. I am happy to be the author of most of what is in here (and, for those parts for which I am not, I can still offer good excuses). But I have taken the chance to strike out certain passages that I now reckon redundant, outdated, or wrong.

My constant companion in this work has been my wife, Ayla. My constant distraction has been the two rascals to whom this book is dedicated, Ada and Samuel. Daniel Dennett at Tufts University, Arien Mack and Judy Friedlander at the New School, and Max Steuer at the LSE have been good friends and critics at every stage. Shelley Cox, of Oxford University Press, and Angela Blackburn have been editors beyond compare.

Contents

1. On Taking Another Look   1

SELVES

2. One Self: A Meditation on the Unity of
   Consciousness   7
3. What Is Your Substance, Whereof Are You Made?   15
4. Speaking for Our Selves: An Assessment of
   Multiple Personality Disorder   19
5. Love Knots   49
6. Varieties of Altruism—and the Common Ground
   Between Them   52

FEELINGS

7. The Uses of Consciousness   65
8. Farewell, Thou Art Too Dear for My Possessing   86
9. How to Solve the Mind–Body Problem   90
10. The Privatization of Sensation   115

DISCOVERIES

11. Mind in Nature   129
12. Cave Art, Autism, and the Evolution of the
    Human Mind   132
13. Scientific Shakespeare   162
14. The Deformed Transformed   165

x

Contents

PRETENCES

15. Tall Stories from Little Acorns Grow 203
16. Behold the Man: Human Nature and
    Supernatural Belief 206
17. Hello, Aquarius! 232
18. Bugs and Beasts Before the Law 235
19. Great Expectations: The Evolutionary Psychology
    of Faith Healing and the Placebo Effect 255

SEDUCTIONS

20. What Shall We Tell the Children? 289
21. The Number of the Beast 318
22. Arms and the Man 320
23. Death in Tripoli 327
24. Follow My Leader 330

Notes 340
Index 363

1

On Taking Another Look

'How often have I said to you,' Sherlock Holmes observed to Dr Watson, 'that when you have eliminated the impossible, whatever remains, *however improbable*, must be the truth?'[1] And how often do we need to be reminded that this is a maxim that is quite generally ignored by human beings?

Here is an immediate test to prove the point. Figure 1 shows a photograph of a strange object, created some years ago in the laboratory of Professor Richard Gregory.[2] What do you see this as being a picture of? What explanation does your mind construct for the data arriving at your eyes?

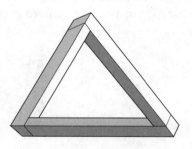

Fig. 1

You see it, presumably, as a picture of the so-called 'impossible triangle': that is, as a picture of a solid triangular object whose parts work perfectly in isolation from one another but whose whole refuses to add up—an object that could not possibly exist in ordinary three-dimensional space.

Yet the fact is that the object in the picture *does* exist in ordinary space. The picture is based on an unretouched photograph of a real object, taken from life, with no kind of optical trickery involved. Indeed, if you were to have been positioned where the camera was at the moment the shutter clicked, you would have seen the real object exactly as you are seeing it on the page.

What, then, should be your attitude to this apparent paradox? Should you perhaps (with an open mind, trusting your personal experience) believe what you unquestionably see, accept that what you always thought could not exist actually does exist, and abandon your long-standing assumptions about the structure of the 'normal' world? Or, taking heed of Holmes's dictum, would you do better instead to make a principled stand against impossibility and go in search of the improbable?

The answer, of course, is that you should do the second. For the fact is that Gregory, far from creating some kind of paranormal object that defies the rules of 3-D space, has merely created a perfectly normal object that defies the rules of human expectation. The true shape of Gregory's 'improbable triangle' is revealed from another camera position in Figure 2.

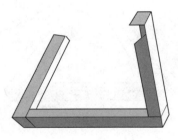

Fig. 2

It is, as it turns out, a most unusual object (there may be only a couple of such objects in existence in the universe). And it has been photographed for Figure 1 from a most unusual point of view (to get this first picture, the camera has had to be placed at the one-and-only position from which the object looks like this). But there it is. And now that you have seen the true solution, presumably you will no longer be taken in.

If only it were so! You look at Figure 2. And now you look back at Figure 1. What do you see this time around? Almost certainly, you still see exactly what you saw before: the impossibility rather than the improbability! Even when prompted in the right direction, you happily, almost casually, continue to 'make sense' of the data in a nonsensical way. Your mind, it seems, cannot help choosing the attractively simple—even if mad—interpretation over the unattractively complicated— even if sane—one. Logic and common sense are being made to play second fiddle to a perceptual ideal of wholeness and completion.

There are many examples in the wider world of human politics and culture where something similar happens: that is to say, where common sense gets overridden by some kind of seductively simple explanatory principle—ethical, political, religious, or even scientific. For, if there is one thing that human beings are amazingly prone to (perhaps we might say good at), it is in emulating the camera operator who took the photograph of Figure 1 and manoeuvring themselves into just the one ideological position from which an impossible, even absurd, explanation of the 'facts of life' happens to look attractively simple and robust.

This special position may be called, for example, Christianity, or Marxism, or Nationalism, or Psychoanalysis—maybe even some forms of science, or scientism. It may be an ideological position that appeals only to some of the human population some of the time or one that appeals to all of the population all of the time. But whichever it is, to those people who, in relation to a particular problem, are currently emplaced in this position, this will almost certainly seem to be the only reasonable place there is to be. 'Here I stand,' in the words of Martin Luther; 'I can do no other.' And the absolute rightness of the stance will seem to be confirmed by the very fact that it permits the welcome solution to the problem that it does.

Yet the telltale sign of what is happening will always be that the solution works only from this one position, and that if the observer were able to shift perspective, even slightly, the gaps in the explanation would appear. Of course, the trick—for those who want to keep faith and save appearances—is not to shift position, or to pull rapidly back if ever tempted.

The lesson is that when would-be gurus offer us final answers to any of life's puzzles, a way of looking at things that brings everything together, the last word on 'how things are', we should be watchful. By all means, let us say: 'Thank you, it makes a pretty picture.' But we should always be prepared to take another look.

SELVES

2

One Self: A Meditation on the Unity of Consciousness

I am looking at my baby son as he thrashes around in his crib, two arms flailing, hands grasping randomly, legs kicking the air, head and eyes turning this way and that, a smile followed by a grimace crossing his face . . . And I'm wondering: what is it like to be him? What is he feeling now? What kind of experience is he having of *himself*?

Then a strong image comes to me. I am standing now, not at the rail of a crib, but in a concert hall at the rail of the gallery, watching as the orchestra assembles. The players are arriving, section by section—strings, percussion, woodwind—taking their separate places on the stage. They pay little if any attention to each other. Each adjusts his chair, smoothes his clothes, arranges the score on the rack in front of him. One by one they start to tune their instruments. The cellist draws his bow darkly across the strings, cocks his head as if savouring the resonance, and slightly twists the screw. The harpist leans into the body of her harp, runs her fingers trippingly along a scale, relaxes and looks satisfied. The oboist pipes a few liquid notes, stops, fiddles with the reed and tries again. The tympanist beats a brief rally on his drum. Each is, for the moment, entirely in his own world, playing only to and for himself, oblivious to anything but his own action and his own sound. The noise from the stage is a medley of single notes and snatches of melody, out of time, out of harmony. Who would believe that all these independent voices will soon be working in concert under one conductor to create a single symphony.

Now, back in the nursery, I seem to be seeing another kind

of orchestra assembling. It is as if, with this baby, all the separate agencies of which he is composed still have to settle into place and do *their* tuning up: nerves need tightening and balancing, sense organs calibrating, pipes clearing, airways opening, a whole range of tricks and minor routines have to be practised and made right. The subsystems that will one day be a system have as yet hardly begun to acknowledge one another, let alone to work together for one common purpose. And as for the conductor who one day will be leading all these parts in concert into life's *Magnificat*: he is still nowhere to be seen.

I return to my question: what kind of experience is this baby having of himself? But, as I ask it, I realize I do not like the answer that suggests itself. If there is no conductor inside him yet, perhaps there is in fact no self yet, and if no self perhaps no experience either—perhaps nothing at all.

If I close my eyes and try to think like a hard-headed philosophical sceptic, I can almost persuade myself it could be so. I must agree that, in theory, there could be no kind of consciousness within this little body, no inner life, nobody at home to have an inner life. But then, as I open my eyes and look at him again, any such scepticism melts. *Someone* in there is surely looking back at me, someone is smiling, someone seems to know my face, someone is reaching out his tiny hand . . . Philosophers think one way, but fathers think another. I can hardly doubt sensations are registering inside this boy, willed actions initiating, memories coming to the surface. However disorganized his life may be, he is surely not totally unconscious.

Yet I realize I cannot leave it there. If these experiences are occurring in the baby boy, they presumably have to belong to an *experiencer*. Every experience has to have a corresponding subject whose experience it is. The point was well made by the philosopher Gottlob Frege, a hundred years ago: it would be absurd, he wrote, to suppose 'that a pain, a mood, a wish should rove about the world without a bearer, independently. An experience is impossible without an experient. The inner world presupposes the person whose inner world it is.'[1]

But if that is the case, I wonder what to make of it. For it seems to imply that those 'someones' that I recognize inside this boy—the someone who is looking, the someone who is acting, the someone who is remembering—must all be genuine subjects of experience (subjects; note the plural). If indeed he does not yet possess a single Self—that Self with a capital S which will later mould the whole system into one — then perhaps he must in fact possess a set of relatively independent sub-selves, each of which must be counted a separate centre of subjectivity, a separate experiencer. Not yet being one person, perhaps he is in fact *many*.

But, isn't this idea bizarre? A lot of independent experiencers? Or—to be clear about what this has to mean—a lot of independent consciousnesses? And all within one body? I confess I find it hard to see how it would work. I try to imagine what it would be like for me to be fractionated in this way and I simply cannot make sense of the idea.

Now, I agree that I myself have many kinds of 'lesser self' inside me: I can, if I try, distinguish a part of me that is seeing, a part that is smelling, a part raising my arm, a part recalling what day it is, and so on. These are certainly different types of mental activity, involving different categories of subjective experience, and I am sure they can properly be said to involve different dimensions of my Self.

I can even agree that these parts of me are a relatively loose confederation that do not all have to be present at one time. Parts of my mind can and do sometimes wander, get lost, and return. When I have come round from a deep sleep, for example, I think it is even true that I have found myself having to gather myself together—which is to say *my selves* together—piecemeal.

Marcel Proust, in *À la recherche du temps perdu*, provides a nice description of just this peculiar experience: 'When I used to wake up in the middle of the night,' he writes,

not knowing where I was, I could not even be sure at first who I was; I had only the most rudimentary sense of existence, such as may lurk

and flicker in the depths of an animal's consciousness . . . But then . . .
out of a blurred glimpse of oil-lamps, of shirts with turned- down
collars, [I] would gradually piece together the original components
of my ego.[2]

So it is true, if I think about this further, that the idea of
someone's consciousness being dispersed in different places is
not completely unfamiliar to me. And yet I can see that this
kind of example will hardly do to help me understand the
baby. For what distinguishes my case from the baby's is precisely that these 'parts of me' that separate and recombine do
not, while separate, exist as distinct and self-sufficient subjects
of experience. When I come together on waking, it is surely
not a matter of my bringing together various sub-selves that
are already separately conscious. Rather, these sub-selves only
come back into existence as and when I plug them back, as it
were, into the main me.

As I stand at the crib watching my baby boy, trying to find
the right way in, I now realize I am up against an imaginative
barrier. I will not say that, merely because I can't imagine it, it
could make no sense at all to suppose that this baby has got all
those separate conscious selves within him. But I will say I do
not know what to say next.

Yet, I am beginning to think there is the germ of some real
insight here. Perhaps the reason why I cannot imagine the
baby's case is tied into that very phrase, '*I* can't imagine . . . '.
Indeed, as soon as I try to imagine the baby as split into several different selves, I make him back into one again by virtue of
imagining it. I imagine each set of experiences as *my* experiences—but, just to the extent that they are all *mine*, they are
no longer separate.

And doesn't this throw direct light on what may be the
essential difference between my case and the baby's? For
doesn't it suggest that it is all a matter of how a person's
experiences are owned—to whom they belong ?

With *me* it seems quite clear that every experience that any
of my sub-selves has is *mine*. And, to paraphrase Frege, in my

case it would certainly make no sense to suppose that a pain, a mood, a wish should rove about my inner world without the bearer in every case being *me*! But maybe with the baby every experience that any of his sub-selves has is not yet *his*. And maybe in his case it does make perfect sense to suppose that a pain, a mood, a wish should rove about inside his inner world without the bearer in every case being *him*.

How so? What kind of concept of 'belonging' can this be, such that I can seriously suggest that, while my experiences belong to me, the baby's do not belong to him? I think I know the answer intuitively; yet I need to work it through.

Let me return to the image of the orchestra. In their case, I certainly want to say that the players who arrive on stage as isolated individuals come to belong to a single orchestra. As an example of 'belonging', this seems as clear as any. But, if there is indeed something that binds the players to belong together, what kind of something is this?

The obvious answer would seem to be the one I have hinted at already: that there is a 'conductor'. After each player settles in and has his period of free play, a dominant authority mounts the stage, lifts his baton, and proceeds to take overall control. Yet, now I am beginning to realize that this image of the conductor as 'chief self' is not the one I want—nor, in fact, was it a good or helpful image to begin with.

Ask any orchestral player, and he'll tell you: although it may perhaps look to an outsider as if the conductor is totally in charge, in reality he often has a quite minor—even a purely decorative—role. Sure, he can provide a common reference point to assist the players with the timing and punctuation of their playing. And he can certainly influence the overall style and interpretation of a work. But that is not what gets the players to belong together. What truly binds them into one organic unit and creates the flow between them is something much deeper and more magical, namely, the very act of making music: that they are together creating a single work of art.

Doesn't this suggest a criterion for 'belonging' that should be much more widely applicable: that parts come to belong to a whole just in so far as they are *participants in a common project*?

Try the definition where you like: what makes the parts of an oak tree belong together—the branches, roots, leaves, acorns ? They share a common interest in the tree's survival. What makes the parts of a complex machine like an aeroplane belong to the aeroplane—the wings, the jet engines, the radar? They participate in the common enterprise of flying.

Then, here's the question: what makes the parts of a person belong together—if and when they do? The clear answer has to be that the parts will and do belong together *just in so far as they are involved in the common project of creating that person's life*.

This, then, is the definition I was looking for. And, as I try it, I immediately see how it works in my own case. I may indeed be made up of many separate sub-selves, but these selves have come to belong together as the one Self that I am because they are engaged in one and the same enterprise: the enterprise of steering me—body and soul—through the physical and social world. Within this larger enterprise each of my selves may indeed be doing its own thing: providing me with sensory information, with intelligence, with past knowledge, goals, judgements, initiatives, and so on. But the point—the wonderful point—is that each self doing its own thing shares a final common path with all the other selves doing their own things. And it is for this reason that these selves are all *mine*, and for this reason that their experiences are all *my experiences*. In short, my selves have become co-conscious through collaboration.

But the baby? Look at him again. There he is, thrashing about. The difference between him and me is precisely that he has as yet no common project to unite the selves within him. Look at him. See how he has hardly started to do anything for himself as a whole: how he is still completely helpless, needy, dependent—reliant on the projects of other people for his survival. Of course, his selves are beginning to get into shape and function on their own. But they do not yet share a final common path. And it is for that reason his selves are not yet all of them *his*, and for that reason their experiences are not yet *his* experiences. His selves are not co-conscious because there is as yet no co-laboration.

Even as I watch, however, I can see things changing. I realize the baby boy is beginning to come together. Already there are hints of small collaborative projects getting under way: his eyes and his hands working together, his face and his voice, his mouth and his tummy. As time goes by, some of these mini-projects will succeed; others will be abandoned. But inexorably over days and weeks and months he will become one coordinated, centrally conscious human being. And, as I anticipate this happening, I begin to understand how in fact he may be going to achieve this miracle of unification. It will not be, as I might have thought earlier, through the power of a supervisory Self who emerges from nowhere and takes control, but through the power inherent in all his sub-selves for, literally, their own *self-organization*.

Then, stand with me again at the rail of the orchestra, watching those instrumental players tuning up. The conductor has not come yet, and maybe he is not ever going to come. But it hardly matters: for the truth is, *it is of the nature of these players to play.* See, one or two of them are already beginning to strike up, to experiment with half-formed melodies, to hear how they sound for themselves, and—remarkably—to find and recreate *their* sound in the *group sound* that is beginning to arise around them. See how several little alliances are forming, the strings are coming into register, and the same is happening with the oboes and the clarinets . See, now, how they are joining together across different sections, how larger structures are emerging.

Perhaps I can offer a better picture still. Imagine, at the back of the stage, above the orchestra, a lone dancer. He is the image of Nijinsky in *The Rite of Spring*. His movements are being shaped by the sounds of the instruments, his body absorbing and translating everything he hears. At first his dance seems graceless and chaotic. His body cannot make one dance of thirty different tunes. Yet, something is changing. See how each of the instrumental players is watching the dancer—looking to find how, within the chaos of those body movements, the dancer is dancing to his tune. And each player, it seems, now wants the dancer to be *his*, to have the dancer give form to *his* sound. But see how, in order to achieve this, each must take account of all the other influences to which the dancer is responding—how each must accommodate to and join in harmony with the entire group. See, then, how, at last, this group of players is becoming *one orchestra* reflected in *the one body of the dancer*—and how the music they are making and the dance that he is dancing have indeed become a *single work of art*.

And my boy, Samuel? His body has already begun to dance to the sounds of his own selves. Soon enough, as these selves come together in creating him, he too will become a single, self-made human being.

3

What Is Your Substance, Whereof Are You Made?

What is your substance, whereof are you made,
That millions of strange shadows on you tend?
Since every one hath, every one, one shade,
And you, but one, can every shadow lend.
Describe Adonis, and the counterfeit,
Is poorly imitated after you;
On Helen's cheek all art of beauty set,
And you in Grecian tires are painted new.
Speak of the spring, and foison of the year;
The one doth shadow of your beauty show,
The other as your bounty doth appear,
And you in every blessèd shape we know.
    In all external grace you have some part,
    But you like none, none you, for constant heart.

Shakespeare, Sonnet LIII

This is a poem, it has been said, of 'abundant flattery'. But not displeasing, I imagine, to the young Earl of Southampton to whom it was probably addressed. Shakespeare had earlier compared him to a summer's day. Now, for good measure, he tells him he combines the promise of spring and the bounty of autumn. He is not only the most beautiful of men—the very picture of Adonis with whom the goddess Venus fell in love— but the equal of Helen, the most beautiful of women, too. What is he made of? Spring, autumn, Adonis, Helen . . . What a piece of work he must have been!

    Yes, what a piece of work. We read the poem, at first

encounter, as if Shakespeare meant it as a straightforward tribute to his friend. But there is not much in Shakespeare's sonnets which is straightforward, and scarcely a line which is simply what it seems.

He is clever, Shakespeare. And he expects us to be clever too. In the very first phrases he hints at a more subtle interpretation of the poem. 'What is your substance, whereof are you made, / that millions of strange shadows on you tend?' Substance and shadow. The reference is to a famous parable by Plato, the story of 'The Cave'.

Imagine, Plato suggests in *The Republic*, that we are in a cave with a great fire burning behind us, whose light casts on the wall our shadows and the shadows of everything else around us. We are chained there facing the wall, unable to look round. We see those dancing shadows, we see life passing in outline before our eyes, but we have no knowledge of the solid reality that lies behind. And so—like a child whose only experience of the world comes through watching a television screen—we come to believe that the shadows themselves are the real thing.

The problem for all of us, Plato implies, is to recognize that beneath the surface of appearances there may exist another level of reality. We see a thing now in this light, now in that. We hear a poem read with this emphasis or that. But every example that reaches our senses is at best an ephemeral and patchy copy, a shadow—and one shadow only—of the transcendental reality behind. Everything and everyone 'has, every one, one shade', and none can reveal all aspects of itself at once.

Except, it seems, for Shakespeare's friend: a being who, contrary both to philosophy and natural law, *can* 'every shadow lend'. A man seen simultaneously by the light of a hundred fires. A poem read in a hundred ways. A shocking new form of a man. As if, among the cut-out portraits, the flat silhouettes which line our shadow-theatre wall, we were to come across a Cubist painting—a portrait by Picasso or by Braque.

Early last century the Cubist painters attempted quite deliberately to overcome the limitations of a single point of view. They took, say, a familiar object—a guitar—broke it apart and portrayed it on the canvas as if it were being seen from several different sides. Calculatedly lending to the object 'every shadow', they hoped that the essence, the inner substance of the object, would shine through. They took a human face and did the same.

But theirs was only the most literal—perhaps the most brutal—attempt to break the chains of Plato's cave. If the Cubists knew the problem, Leonardo da Vinci knew it too. If the Cubists solved it by superimposing different points of view, so in another sense did Leonardo. If a hundred shadows tend upon Braque's *Lady with Guitar*, surely a thousand tend upon the *Mona Lisa*.

'Hers is the head upon which "all the ends of the world are come"'—the words are the critic Walter Pater's.

All the thoughts and experience of the world are etched and moulded there: the animalism of Greece, the lust of Rome, the reverie of the middle age with its spiritual ambition and imaginative loves, the return of the Pagan world, the sins of the Borgias. She is older than the rocks among which she sits; like the vampire, she has been dead many times, and learned the secrets of the grave; and has been a diver in deep seas, and keeps their fallen days about her; and trafficked for strange webs with Eastern merchants; and, as Leda, was the mother of Helen of Troy, and, as Saint Anne, the mother of Mary.[1]

Leonardo, in making a portrait of a Florentine merchant's wife, has given us a picture of someone with the power to represent all ways of existing to all men. Now Shakespeare, in writing to an Elizabethan youth, tells us of just such another one.

And tells us, in other poems, that he is in love with him. Alas, poor Shakespeare! To love, and to ask for love, from such a complicated being—or rather such a complicated nest of different beings is as foolish as to try to play a simple serenade on Braque's guitar.

Such charismatic prodigies—the Mona Lisa, Shakespeare's

friend, or, in our own time, Nijinsky, Greta Garbo, Marilyn Monroe—may have an almost magical attraction. But the spell they exert is the spell that Stanislavski called 'stage charm'. By face, by manner, by voice and personality they hint at every possible existence; but never do they finally confirm or deny a single one of them. All things to all people, but never at last the capacity to be any particular thing to anyone.

Such men—such women too, of course—are dangerous. So dangerous that Plato himself in describing his ideal Republic recommended that we ban them from the city gates. And if Shakespeare knew Plato's story of the cave, he also very likely knew of Plato's warning. 'If any such man,' Plato writes, 'who has the skill to transform himself into all sorts of characters and represent all sorts of things, shall come to us to show us his art, we shall anoint him with myrrh and set a garland of wool upon his head . . . and send him away to another city.'[2]

Shakespeare, in the other sonnets, chides, rails, complains against his friend. But at no time has he the courage to send the man away.

Yet Shakespeare knew—none better—the trouble he was in. The poem in its deeper meaning is no poem of 'abundant flattery'. It's a lament.

4

Speaking for Our Selves: An Assessment of Multiple Personality Disorder

NICHOLAS HUMPHREY AND DANIEL C. DENNETT

> Thus play I in one person many people, and none contented.
>
> Shakespeare, *Richard II*, V. v. 31–2

In the early 1960s, when the laws of England allowed nudity on stage only if the actor did not move, a tent at the Midsummer Fair in Cambridge offered an interesting display. 'The one and only Chamaeleon Lady,' the poster read, 'becomes Great Women in History.' The inside of the tent was dark. 'Florence Nightingale!' the showman bellowed, and the lights came up on a naked woman, motionless as marble, holding up a lamp. The audience cheered. The lights went down. There was a moment's shuffling on the stage. 'Joan of Arc!' and here she was, lit from a different angle, leaning on a sword. 'Good Queen Bess!' and now she had on a red wig and was carrying an orb and sceptre . . . 'But it's the same person,' said a know-all schoolboy.

Imagine now, thirty years later, a commercial for an IBM computer. A poster on a tent announces, 'The one and only IBM PC becomes Great Information Processors of History'. The tent is dark. 'WordStar!' shouts the showman, and the lights come up on a desktop computer, displaying a characteristic menu of commands. The lights go down. There is the sound of changing disks. 'Paintbrush!' and here is the computer displaying a different menu. 'Now, what you've all been waiting for, Lotus 123!' . . . 'But it's just a different program,' says the schoolboy.

Somewhere between these two scenarios lies the phenomenon of multiple personality in human beings. And
somewhere between these two over-easy assessments of
it lie we. One of us (NH) is a theoretical psychologist, the
other (DCD) is a philosopher, both with a long- standing
interest in the nature of personhood and of the self. We have
had the opportunity during the past year to meet several
'multiples', to talk with their therapists, and to savour the
world from which they come. We give here an outsider's inside
view.

We had been at the conference on Multiple Personality Disorder for two full days before someone made the inevitable
joke: 'The problem with those who don't believe in MPD is
they've got Single Personality Disorder.' In the mirror-world
that we had entered, almost no one laughed.

The occasion was the Fifth International Conference on
Multiple Personality/Dissociative States in Chicago in
October 1988, attended by upwards of five hundred psy-
chotherapists and a large but unquantifiable number of former patients.

The Movement or the Cause (as it was called) of MPD has
been undergoing an exponential growth: 200 cases of multiplicity reported up till 1980, 1,000 known to be in treatment
by 1984, 4,000 now. Women outnumber men by at least four
to one, and there is reason to believe that the vast majority—
perhaps 95 per cent—have been sexually or physically abused
as children. We heard it said there are currently more than
25,000 multiples in North America.[1]

The accolade of 'official diagnosis' was granted in 1980,
with an entry in the clinician's handbook, *DSM-III*:[2]

Multiple Personality. 1. The existence within an individual of two
or more distinct personalities, each of which is dominant at a particular time. 2. The personality that is dominant at any particular time
determines the individual's behavior. 3. Each individual personality is complex and integrated with its own unique behavior patterns
and social relationships.

Typically there is said to exist a 'host' personality, and several alternative personalities or 'alters'. Usually, though not always, these personalities call themselves by different names. They may talk with different accents, dress by choice in different clothes, frequent different locales.

None of the personalities is emotionally well rounded. The host is often emotionally flat, and different alters express exaggerated moods: Anger, Nurturance, Childishness, Sexiness. Because of their different affective competence, it falls to different alters to handle different social situations. Thus one may come out for love-making, another for playing with the kids, another for picking a fight, and so on.

The host personality is on stage most of the time, but the alters cut in and displace the host when for one reason or another the host cannot cope. The host is usually amnesic for those episodes when an alter is in charge; hence the host is likely to have blank spots or missing time. Although general knowledge is shared between them, particular memories are not.

The life experience of each alter is formed primarily by the episodes when she or he is in control. Over time, and many episodes, this experience is aggregated into a discordant view of who he or she is—and hence a separate sense of self.

The number of alters varies greatly between patients, from just one (dual personality) to several dozen. In the early literature most patients were reported to have two or three, but there has been a steady increase, with a recent survey suggesting the median number is eleven. When the family has grown this large, one or more of the alters is likely to claim to be of different gender.

Such at least is how we first heard multiplicity described to us. It was not, however, until we were exposed to particular case histories, that we ourselves began to have any feeling for the human texture of the syndrome or for the analysis being put on it by MPD professionals. Each case must be, of course, unique. But it is clear that common themes are beginning to emerge, and that, based on their pooled experience, therapists are beginning to think in terms of a 'typical case history'.[3] The case that follows, although in part a reconstruction, is true to type (and life).

Mary, in her early thirties, has been suffering from depression, confusional states, and lapses of memory. During the last few years she has been in and out of the hospital, where she has been diagnosed variously as schizophrenic, borderline, and manic depressive. Failing to respond to any kind of drug treatment, she has also been suspected of malingering. She ends up eventually in the hands of Dr R, who specializes in treating dissociative disorders. More trusting of him than of previous doctors, Mary comes out with the following telltale information.

Mary's father died when she was two years old, and her mother almost immediately remarried. Her stepfather, she says, was kind to her, although 'he sometimes went too far'. Through childhood she suffered from sick headaches. She had a poor appetite and she remembers frequently being punished for not finishing her food. Her teenage years were stormy, with dramatic swings in mood. She vaguely recalls being suspended from her high school for a misdemeanour, but her memory for her school years is patchy. In describing them she occasionally resorts—without notice—to the third person ('She did this . . . That happened to her'), or sometimes the first person plural ('We [Mary] went to Grandma's'). She is well informed in many areas, is artistically creative, and can play the guitar; but when asked where she learnt it, she says she does not know and deflects attention to something else. She agrees that she is 'absent-minded'—'but aren't we all?': for example, she might find there are clothes in her closet that she can't remember buying, or she might find she has sent her niece two birthday cards. She claims to have strong moral values; but other people, she admits, call her a hypocrite and liar. She keeps a diary—'to keep up', she says, 'with where we're at'.

Dr R (who already has four multiples in treatment) is beginning to recognize a pattern. When, some months into treatment, he sees Mary's diary and observes that the handwriting varies from one entry to the next, as if written by several different people, he decides (in his own words) 'to go for gold'. With Mary's agreement, he suggests they should undertake an exploratory session of hypnosis. He puts her into a light trance and requests that the 'part of Mary that hasn't yet come forward' should make herself known. A sea-change occurs in the woman in front of him. Mary, until then a model of decorum, throws him a flirtatious smile. 'Hi, Doctor,' she says, 'I'm Sally . . . Mary's a wimp. She thinks she knows it all, but I can tell you . . .'

But Sally does not tell him much, at least not yet. In subsequent sessions (conducted now without hypnosis) Sally comes and goes, almost as if she were playing games with Dr R. She allows him glimpses of what she calls the 'happy hours', and hints at having a separate and exotic history unknown to Mary. But then with a toss of the head she slips away—leaving Mary, apparently no party to the foregoing conversation, to explain where she has been.

Now Dr R starts seeing his patient twice a week, for sessions that are several hours in length. In the course of the next year he uncovers the existence not just of Sally but of a whole family of alter personalities, each with their own characteristic style. 'Sally' is coquettish, 'Hatey' is angry, 'Peggy' is young and malleable. Each has a story to tell about the times when she is 'out in front'; and each has her own set of special memories. While each of the alters claims to know most of what goes on in Mary's life, Mary herself denies anything but hearsay knowledge of their roles.

To begin with, the change-over from one personality to another is unpredictable and apparently spontaneous. The only clue that a switch is imminent is a sudden look of vacancy, marked perhaps by Mary's rubbing her brow, or covering her eyes with her hand (as if in momentary pain). But as their confidence grows, it becomes easier for Dr R to summon different alters 'on demand'.

Dr R's goal for Mary now becomes that of 'integration'—a fusing of the different personalities into one self. To achieve this he has not only to acquaint the different alters with each other, but also to probe the origins of the disorder. Thus he presses slowly for more information about the circumstances that led to Mary's 'splitting'. Piecing together the evidence from every side, he arrives at—or is forced to—a version of events that he has already partly guessed. This is the story that Mary and the others eventually agree upon:

When Mary was four years old, her stepfather started to take her into his bed. He gave her the pet name Sandra, and told her that 'Daddy-love' was to be Sandra's and his little secret. He caressed her and asked for her caresses. He ejaculated against her tummy. He did it in her bottom and her mouth. Sometimes Mary tried to please him. Sometimes she lay still like a doll. Sometimes she was sick and cried that she could take no more. One time she said that she would tell—but the man hit her and said that both of them would go to prison. Eventually, when the pain, dirt, and disgrace became too much to bear, Mary simply 'left it all behind': while the man abused her, she dissociated and took off to another world. She left—and left Sandra in her place.

What happened next is, Dr R insists, no more than speculation. But he pictures the development as follows. During the next few crucial years—those years when a child typically puts down roots into the fabric of human society, and develops a unitary sense of 'I' and 'Me'—Mary was able to function quite effectively. Protected from all knowledge of the horror, she had a comprehensible history, comprehensible feelings, and comprehensible relationships with members of her family. The 'Mary-person' that she was becoming was one person with one story.

Mary's gain was, however, Sandra's loss. For Sandra knew. And this knowledge, in the early years, was crippling. Try as she might, there was no single story that she could tell that would embrace her contradictory experiences; no one 'Sandra-person' for her to become. So Sandra, in a state of inchoateness, retreated to the shadows, while Mary—except for 'Daddy- love'—stayed out front.

Yet if Mary could split, then so could Sandra. And such, it seems, is what occurred. Unable to make it all make sense, Sandra made sense from the pieces—not consciously and deliberately, of course, but with the cunning of unconscious design: she parcelled out the different aspects of her abuse experience, and assigned each aspect to a different self (grafting, as it were, each set of memories as a side-branch to the existing stock she shared with Mary). Thus her experience of liking to please Daddy gave rise to what became the Sally-self. Her experience of the pain and anger gave rise to Hatey. And her experience of playing at being a doll gave rise to Peggy.

Now these descendants of the original Sandra could, with relative safety, come out into the open. And before long, opportunities arose for them to try their new-found strength in settings other than that of the original abuse. When Mary lost her temper with her mother, Hatey could chip in to do the screaming. When Mary was kissed by a boy in the playground, Sally could kiss him back. Everyone could do what they were 'good at'—and Mary's own life was made that much simpler. This pattern of what might be termed 'the division of emotional labour' or 'self-replacement therapy' proved not only to be viable, but to be rewarding all around.

Subsequently this became the habitual way of life. Over time, each member of the family progressively built up her own separate store of memories, competencies, idiosyncrasies, and social styles. But they were living in a branching house of cards. During her teenage years, Mary's varying moods and waywardness could be passed off as 'adolescent rebelliousness'. But in her late twenties, her true fragility began to show—and she lapsed into confusion and depression.

Although we have told this story in what amounts to cartoon form, we have no doubts that cases like Mary's are authentic. Or, rather, we should say we have no doubts that there are real people and real doctors to whom this case history could very well apply. Yet—like many others who have taken a sceptical position about MPD—we ourselves have reservations about what such a case history in fact amounts to.

How could anyone know for sure the events were as described? Is there independent confirmation that Mary was abused? Does her story match with what other people say about her? How do we know the whole thing is not just an hysterical invention? To what extent did the doctor lead her on? What transpired during the sessions of hypnosis? And, anyway, what does it all really mean? What should we make of Dr R's interpretation? Is it really possible for a single human being to have several different 'selves'?

The last problem—that of providing a philosophically and scientifically acceptable theory of MPD—is the one we have a special interest in addressing. You might think, however, we ought to start with a discussion of the 'factual evidence': for why discuss the theoretical basis of something that has not yet been proven to exist? Our answer is that unless and until MPD can be shown to be theoretically possible—that is, to be neither a logical nor a scientific contradiction—any discussion of the evidence is likely to be compromised by a priori disbelief.

As Hume remarked in his essay 'Of Miracles': 'it is a general maxim worthy of our attention . . . that no testimony is sufficient to establish a miracle unless the testimony be of such a kind that its falsehood would be more miraculous than the fact which it endeavours to establish.'[4] In the history of science there have been many occasions in which seemingly miraculous phenomena were not and perhaps could not be taken seriously until some form of theoretical permission for them had been devised (the claims of acupuncture, for example, were assumed by Western scientists to make no sense—and hence to be false—until the discovery of endogenous opiates paved the way for a scientific explanation). We shall, we hope, be in a better position to assess the testimony concerning MPD— that is, to be both critical and generous—if we can first make a case that the phenomenon is not only possible but even (in certain circumstances) plausible.

Many people who find it convenient or compelling to talk about the 'self' would prefer not to be asked the emperor's-new-clothes question: just what, exactly, is a 'self'? When confronted by an issue that seems embarrassingly metaphysical, it is tempting to temporize and wave one's hands: 'It's not a thing, exactly, but more a sort of, well, a concept or an organizing principle or . . .' This will not do. And yet what will?

Two extreme views can be and have been taken. Ask a layman what he thinks a self is, and his unreflecting answer will probably be that a person's self is indeed some kind of real thing: a ghostly supervisor who lives inside his head, the thinker of his thoughts, the repository of his memories, the holder of his values, his conscious inner 'I'. Although he might be unlikely these days to use the term 'soul', it would be very much the age-old conception of the soul that he would have in mind. A self (or soul) is an existent entity with executive powers over the body and its own enduring qualities. Let's call this realist picture of the self, the idea of a 'proper-self'.

Contrast it, however, with the revisionist picture of the self which has become popular among certain psychoanalysts and philosophers of mind. On this view, selves are not things at all, but instead are explanatory fictions. Nobody really has a soul-like agency inside them: we just find it useful to imagine the existence of this conscious inner 'I' when we try to account for their behaviour (and, in our own case, our private stream of consciousness). We might say indeed that the self is rather like the 'centre of narrative gravity' of a set of biographical events and tendencies; but, as with a centre of physical gravity, there's really no such thing (with mass or shape or colour).[5] Let's call this non-realist picture of the self, the idea of a 'fictive-self'.

Now maybe (one might think) it is just a matter of the level of description: the plain man's proper-self corresponds to the intrinsic reality, while the philosopher's fictive-selves correspond to people's (necessarily inadequate) attempts to grasp that intrinsic reality. So, for example, there is indeed a proper-Nicholas-Humphrey-self that actually resides inside one of the authors of this essay, and alongside it there are the various fictive-Humphrey-selves that he and his acquaintances have reconstructed: Humphrey as seen by Humphrey, Humphrey as seen by Dennett, Humphrey as seen by Humphrey's mother, and so on.

This suggestion, however, would miss the point of the revisionist critique. The revisionist case is that, to repeat, there really is no proper-self: none of the fictive-Humphrey-selves—including Humphrey's own first-hand version—corresponds to anything that actually exists in Humphrey's head.

At first sight this may not seem reasonable. Granted that whatever is inside the head might be difficult to observe, and granted also that it might be a mistake to talk about a 'ghostly supervisor', nonetheless there surely has to be some kind of a supervisor in there: a supervisory brain program, a central controller, or whatever. How else could anybody function—as most people clearly do function—as a purposeful and relatively well-integrated agent?

The answer that is emerging from both biology and Artificial Intelligence is that complex systems can in fact function in what seems to be a thoroughly 'purposeful and integrated' way simply by having lots of subsystems doing their own thing without any central supervision. Indeed, most systems on earth that appear to have central controllers (and are usefully described as having them) do not. The behaviour of a termite colony provides a wonderful example of this. The colony as a whole builds elaborate mounds, gets to know its territory, organizes foraging expeditions, sends out raiding parties against other colonies, and so on. The group cohesion and coordination is so remarkable that hard-headed observers have been led to postulate the existence of a colony's 'group soul' (see Marais's 'soul of the white ant'). Yet in fact all this group wisdom results from nothing other than myriads of individual termites, specialized as several different castes, going about their individual business—influenced by each other, but quite uninfluenced by any master plan.[6]

Then is the argument between the realists and the revisionists being won hands down by the revisionists? No, not completely. Something (some thing?) is missing here. But the question of what the 'missing something' is, is being hotly debated by cognitive scientists in terms that have become increasingly abstruse. Fortunately we can avoid—maybe even leapfrog—much of the technical discussion by the use of an illustrative metaphor (reminiscent of Plato's *Republic*, but put to quite a different use).

Consider the United States of America. At the fictive level there is surely nothing wrong with personifying the USA and talking about it (rather like the termite colony) as if it had an inner self. The USA has memories, feelings, likes and dislikes, hopes, talents, and so on. It hates Communism, is haunted by the memory of Vietnam, is scientifically creative, socially clumsy, somewhat given to self-righteousness, rather sentimental. But does that mean (here is the revisionist speaking) there is one central agency inside the USA which embodies all those qualities? Of course not. There is, as it happens, a specific area of the country where much of it comes together. But go to Washington and ask to speak to Mr American Self, and you'd find there was nobody home: instead, you'd find a lot of different agencies (the Defense Department, the Treasury, the courts, the Library of Congress, the National Science Foundation, and so on) operating in relative independence of each other.

To be sure (and now it is the realist speaking), there is no such thing as Mr American Self, but as a matter of fact there is in every country on earth a Head of State: a President, Queen, Chancellor, or some such figurehead. The Head of State may actually be non- executive; certainly he does not himself enact all the subsidiary roles (the US President does not bear arms, sit in the courts, play baseball, or travel to the Moon). But nevertheless he is expected at the very least to take an active interest in all these national pursuits. The President is meant to appreciate better than anyone the 'State of the Union'. He is meant to represent different parts of the nation to each other, and to inculcate a common value system. Moreover—and this is most important—he is the 'spokesman' when it comes to dealing with other nation states.

That is not to say that a nation, lacking such a figurehead, would cease to function day-to-day. But it is to say that in the longer term it may function much better if it does have one. Indeed, a good case can be made that nations, unlike termite colonies, require this kind of figurehead as a condition of their political survival—especially given the complexity of international affairs.

The drift of this analogy is obvious. In short, a human being too may need an inner figurehead—especially given the complexities of human social life. Consider, for example, the living body known as Daniel Dennett. If we were to look around inside his brain for a Chief Executive module, with all the various mental properties we attribute to Dennett himself, we would be disappointed. Nonetheless, were we to interact with Dennett on a social plane, both we and he would soon find it essential to recognize someone—some figurehead—as his spokesman and indeed his leader. Thus we come back full circle, though a little lower down, to the idea of a proper-self: not a ghostly supervisor, but something more like a 'Head of Mind' with a real, if limited, causal role to play in representing the person to himself and to the world.[7]

If this is accepted (as we think it should be), we can turn to the vexed question of self- development or self-establishment. Here the Head of State analogy may seem at first less helpful. For one thing, in the USA at least, the President is democratically elected by the population. For another, the candidates for the presidency are pre-formed entities, already waiting in the wings.

Yet is this really so? It could equally be argued that the presidential candidates, rather than being pre-formed, are actually brought into being—through a narrative dialectical process—by the very population to which they offer their services as President. Thus the population (or the news media) first try out various fictive versions of what they think their 'ideal President' should be, and then the candidates adapt themselves as best they can to fill the bill. To the extent that there is more than one dominant fiction about 'what it means to be American', different candidates mould themselves in different ways. But in the end only one can be elected—and he will of course claim to speak for the whole nation.

In very much a parallel way, we suggest, a human being first creates—unconsciously—one or more ideal fictive-selves and then elects the best supported of these into office as her Head of Mind. A significant difference in the human case, however, is that there is likely to be considerably more outside influence. Parents, friends, and even enemies may all contribute to the image of 'what it means to be me', as well as—and maybe over and above—the internal news media. Daddy, for example, might lean on the growing child to impose an invasive fictive-self.

Thus a human being does not start out as single or as multiple; she starts out without any Head of Mind at all. In the normal course of development, she slowly gets acquainted with the various possibilities of selfhood that 'make sense', partly through her own observation, partly through outside influence. In most cases a majority view emerges, strongly favouring one version of 'the real me', and it is that version which is installed as her elected Head of Mind. But in some cases the competing fictive-selves are so equally balanced, or different constituencies within her are so unwilling to accept the result of the election, that constitutional chaos reigns— and there are snap elections (or *coups d'état*) all the time.

Could a model inspired by (underlying, rendering honest) this analogy account for the memory black-spots, differences in style, and other symptomatology of MPD? Certainly the analogy provides a wealth of detail suggesting so. Once in office, a new Head of State typically downplays certain 'unfortunate' aspects of his nation's history (especially those associated with the rival Head of State who immediately preceded

him). Moreover, he himself, by standing for particular national values, affects the course of future history by encouraging
the expression of those values by the population (and so, by a
kind of feedback, confirming his own role).

Let's go back to the case of Mary. As a result of her experience of abuse, she (the whole, disorganized, conglomeration
of parts) came to have several alternative pictures of the real
Mary, each championed by different constituencies within her.
So incompatible were these pictures, yet so strong were the
electoral forces, that there could be no lasting agreement on
who should represent her. For a time the Mary constituency
got its way, overriding the Sandra constituency. But later the
Sandra forces subdivided, to yield Sally, Hatey, Peggy; and
when the opportunities arose, these reformed forces began to
win electoral battles. She became thus constitutionally unstable, with no permanent solution to the question of 'who I
really am'. Each new (temporarily elected) Head of Mind
emphasized different aspects of her experience and blocked
off others; and each brought out exaggerated character traits.

We have talked here in metaphors. But translations into the
terms of current cognitive science would not be difficult to
formulate. First, what sense can be given to the notion of a
'Head of Mind'? The analogy with a spokesman may not be
far off the literal truth. The language-producing systems of the
brain have to get their instructions from somewhere, and the
very demands of pragmatics and grammar would conspire to
confer something like Head of Mind authority on whatever
subsystem currently controls their input. E. M. Forster once
remarked, 'How can I tell what I think until I see what I say?'
The four 'I's in this sentence are meant to refer to the same
thing. But this grammatical tradition may depend—and may
always have depended—on the fact that the thought
expressed in Forster's question is quite literally self-confirming: what 'I' (my self) thinks is what 'I' (my language apparatus) says.

There can, however, be no guarantee that either the speaker or anyone else who hears him over an extended period will

settle on there being just a single 'I'. Suppose, at different times, different subsystems within the brain produce 'clusters' of speech that simply cannot easily be interpreted as the output of a single self. Then—as a Bible scholar may discover when working on the authorship of what is putatively a single-authored text—it may turn out that the clusters make best sense when attributed to different selves.

How about the selective amnesia shown by different Heads of Mind? To readers who have even a passing knowledge of computer information processing, the idea of mutually inaccessible 'directories' of stored information will already be familiar. In cognitive psychology, new discoveries about state-dependent learning and other evidence of modularization in the brain have led people to recognize that failure of access between different subsystems is the norm rather than the exception. Indeed, the old Cartesian picture of the mind 'transparent to itself' now appears to be rarely if ever achievable (or even desirable) in practice. In this context, the out-of-touchness of different selves no longer looks so startling.

What could be the basis for the different 'value systems' associated with rival Heads of Mind? At another level of analysis, psychopharmacological evidence suggests that the characteristic emotional style of different personalities could correspond to the brain-wide activation or inhibition of neural pathways that rely on different neurotransmitter chemicals. Thus the phlegmatic style of Mary's host personality could be associated with low norepinephrine levels, the shift to the carnal style of Sally with high norepinephrine, and the out-of-control Hatey with low dopamine.

Even the idea of an 'election' of the current Head of Mind is not implausible. Events very like elections take place in the brain all the time, whenever coherent patterns of activity compete for control of the same network. Consider what happens, for example, when the visual system receives two conflicting images at the two eyes. First there is an attempt at fusion; but if this proves to be unstable, 'binocular rivalry' results, with the input from one eye completely taking over while the other is suppressed. Thus we already have, at the level of visual neu-rophysiology, clear evidence of the mind's general preference for single-mindedness over completeness.

These ideas about the nature of selves are by no means alto-gether new. C. S. Peirce, for instance, expressed a similar vision in 1905:

A person is not absolutely an individual. His thoughts are what he is 'saying to himself', that is, is saying to that other self that is just com-ing into life in the flow of time.[8]

From within the psychoanalytic tradition, Heinz Kohut wrote:

I feel that a formulation which puts the self into the centre of the per-sonality as the initiator of all actions and as the recipient of all impressions exacts too high a price . . . If we instead put our trust in empirical observation .. we will see different selves, each of them a lasting psychological configuration, . . . fighting for ascendancy, one blocking out the other, forming compromises with each other, and acting inconsistently with each other at the same time. In general, we will witness what appears to be an uneasy victory of one self over all others.[9]

Robert Jay Lifton has defined the self as the 'inclusive symbol of one's own organism'; and in his discussions of what he calls 'proteanism' (an endemic form of multiplicity in modern human beings) and 'doubling' (as in the double life led by Nazi doctors), he has stressed the struggle that all human beings have to keep their rival self-symbols in symbiotic harmony.[10]

These ideas have, however, been formulated without refer-ence to the newly gathered evidence on MPD. Moreover, the emphasis of almost all the earlier work has been on the under-lying continuity of human psychic structure: a single stream of consciousness manifesting itself in now this, now that config-uration. Nothing in the writings of Kohut or of Lifton would have prepared us for the radical discontinuity of conscious-ness that—if it really exists—is manifest in the case of a multiple like Mary.

Which brings us to the question that has been left hanging all along: does 'real MPD' exist? We hope that, in the light of the preceding discussion, we shall be able to come closer to an answer.

What would it mean for MPD to be 'real'? We suggest that, if the model we have outlined is anything like right, it would mean at least the following:

1. The subject will have, at different times, different 'spokespersons', corresponding to separate Heads of Mind. Both objectively and subjectively, this will be tantamount to having different 'selves' because the access each such spokesman will have to the memories, attitudes, and thoughts of other spokespersons will be, in general, as indirect and intermittent as the access one human being can have to the mind of another.

2. Each self, when present, will claim to have conscious control over the subject's behaviour. That is, this self will consider the subject's current actions to be her actions, experiences to be her experiences, memories to be her memories, and so on. (At times the self out front may be conscious of the existence of other selves—she may even hear them talking in the background—but she will not be conscious with them).

3. Each self will be convinced—as it were by 'her own rhetoric'—about her own integrity and personal importance.

4. This self-rhetoric will be convincing not only to the subject but also (other things being equal) to other people with whom she interacts.

5. Different selves will be interestingly different. That is, each will adopt a distinctive style of presentation—which very likely will be associated with differences in physiology.

To which we would add—not necessarily as a criterion of 'real

'multiplicity' but none the less as an important factual issue—
that:

> 6. The 'splitting' into separate selves will generally have
> occurred before the patient entered therapy.

Now, what are the facts about MPD? The first thing to say
is that in no case do we know that all these criteria have been
met. What we have to go on instead is a plethora of isolated
stories, autobiographical accounts, clinical reports, police
records, and just a few scientific studies. Out of those the following answers form.

Does the phenomenon exist?

There can be no doubt that what might be called a 'candidate
phenomenon' exists. There are literally thousands of people
living today who, in the course of clinical investigation, have
presented themselves as having several independent selves (or
'spokespersons' for their minds). Such cases have been
described in reputable scientific journals, recorded on film,
shown on television, cross-examined in law courts. We ourselves have met with several of them and have even argued
with these separate selves about why we should believe the
stories that they tell us. Sceptics may still choose to doubt
what the phenomenon amounts to, but they should no longer
doubt that it occurs.

Do multiples themselves believe in what they are saying?

Certainly they seem to do so. In the clinic, at least, different
selves stoutly insist on their own integrity, and resist any suggestion that they might be 'play-acting' (a suggestion which,
admittedly, most therapists avoid). The impression they make
is not of someone who is acting, but rather of a troubled individual who is doing her best—in what can only be described as
difficult circumstances—to make sense of what she takes to be
the facts of her experience.

As persuasive as anything is the apparently genuine puzzlement that patients show when confronted by facts they can't make sense of. Thus one woman told us of how, when—as frequently happened—she came home and found her neat living room all messed up, she suspected that other people must be playing tricks on her. A young man described how he found himself being laughed at by his friends for having been seen around gay bars: he tried over several months to grow a beard to prove his manhood, but as soon as the stubble began to sprout, someone—he did not know who—shaved it off. A woman discovered that money was being mysteriously drawn from her bank account, and told the police that she was being impersonated. We have heard of a case of a highly sceptical patient who refused to accept her therapist's diagnosis until they both learned that one of her alters was seeing another therapist.

That is not to say that such stories would always stand up to critical examination: examination, that is, by the standards of 'normal human life'. But this, it seems, is quite as much a problem for the patient as for anyone else. These people clearly know as well as anybody that there is something wrong with them and that their lives don't seem to run as smoothly as other people's. In fact it would be astonishing (and grounds for our suspicion) if they did not: for, to coin a phrase, they were not born yesterday, and they are generally too intelligent not to recognise that in some respects their experience is bizarre. We met a woman, Gina, with a male alter, Bruce, and asked Bruce the obvious 'normal' question: when he goes to the bathroom, does he choose the Ladies or the Gents? He confessed that he goes to the Ladies—because 'something went wrong with my anatomy' and 'I turned out to be a male living in a woman's body'.

For several years a multiple newsletter—*S4OS* (*Speaking for Our Selves*)—circulated, in which patients shared with each other their experiences and strategies. In September 1987, *S4OS* claimed 691 subscribers.[11]

*Do they succeed in persuading other people to believe
in them?*

We have no doubt that the therapist who diagnoses MPD is
fully convinced that he is dealing with several different selves.
But, from our standpoint, a more crucial issue is whether
other people who are not already au fait with the diagnosis
accept this way of looking at things. According to our analysis (or indeed any other we can think of), selves have a public
as well as a private role to play: indeed, they exist primarily to
handle social interactions. It would therefore be odd, to say
the least, if some or all of a patient's selves were to be kept
entirely secret from the world.

On this point the evidence is surprisingly patchy. True
enough, in many cases the patient herself will—in the context
of the therapeutic situation—tell stories of her encounters in
the outside world. But what we need is evidence from a third
source: a neutral source that is in no way linked to the context
in which splitting is 'expected' (as might still be the case with
another doctor, or another patient or even a television journalist). We need to know whether the picture of her multiple
life that the therapist and patient have worked out together
jibes with what other people have independently observed.

Prima facie, it sounds like the kind of evidence it would be
easy to obtain—by asking family, friends, workmates, or
whomever. There is the problem, of course, that certain lines
of enquiry are ruled out on ethical grounds, or because their
pursuit would jeopardize the patient's ongoing therapy, or
would simply involve an unjustifiable amount of time.
Nonetheless it is disappointing to discover how few such
enquiries have been made.

Many multiple patients are married and have families;
many have regular employment. Yet, again and again it seems
that no one on the outside has in fact noticed anything peculiar—at least not so peculiar. Maybe, as several therapists
explained to us, their patients are surprisingly good at 'covering up' (secrecy, beginning in childhood, is part and parcel of

the syndrome—and in any case the patient has probably learned to avoid putting herself or others on the spot). Maybe other people have detected something odd and dismissed it as nothing more than inconstancy or unreliability (after all, everyone has changing moods, most people are forgetful, and many people lie). Gina told us of how she started to make love to a man she met at an office party but grew bored with him and left—leaving 'one of the kids' (another alter) cringing in her place. The man, she said, was quite upset. But no one has heard his side of the story.

To be sure, in many cases, perhaps even most, there is some form of post-diagnostic confirmation from outside: the husband who, when the diagnosis is explained to him, exclaims 'Now it all makes sense!', or the boyfriend who volunteers to the therapist tales of what it is like to be 'jerked around' by the tag-team alters of his partner. One patient's husband admitted to mixed emotions about the impending cure or integration of his wife: 'I'll miss the little ones!'

The problem with such retrospective evidence is, however, that the informant may simply be acceding to what might be termed a 'diagnosis of convenience'. It is probably the general rule that once multiplicity has been recognized in therapy, and the alters have been 'given permission' to come out, there are gains to be had all round from adopting the patient's preferred style of presentation. When we ourselves were introduced to a patient who switched three times in the course of half an hour, we were chastened to discover how easily we ourselves fell in with addressing her as if she were now a man, now a woman, now a child—a combination of good manners on our part and an anxiety not to drive the alter personality away (as Peter Pan said, 'Every time someone says "I don't believe in fairies", there is a fairy somewhere who falls down dead').

Any interaction with a patient involves cooperation and respect, which shade imperceptibly into collusion. The alternative might be surreptitious observation in extra-clinical situations, but this would be as hard to justify as to execute. The result is that one is limited to encounters that—in our limited experience—have an inevitable séance-like quality to them.

Therapists with whom we have talked are defensive on this issue. We have to say, however, that, so far as we can gather, evidence for the external social reality of MPD is weak.

Are there 'real' differences between the different selves?

One therapist confided to us that, in his view, it was not uncommon for the different selves belonging to a single patient to be more or less identical—the only thing distinguishing them being their selective memories. More usually, however, the selves are described as being manifestly different in both mental and bodily character. The question is: do such differences go beyond the range of 'normal' acting out?

At the anecdotal level, the evidence is tantalizing. For example, a psychopharmacologist (whom we have reason to consider as hard-headed as they come) told us of how he discovered to his astonishment that a male patient, whose host personality could be sedated with 5mg of valium, had an alter personality who was apparently quite impervious to the drug: the alter remained as lively as ever when given a 50mg intravenous dose (sufficient in most people to produce anaesthesia).

Any would-be objective investigator of MPD is soon struck by the systematic elusiveness of the phenomena. Well-controlled scientific studies are few (and for obvious reasons difficult to do). Nonetheless, what data there are all go to show that multiple patients—in the context of the clinic—may indeed undergo profound psychophysiological changes when they change personality state. There is preliminary evidence, for example, of changes in handedness, voice patterns, evoked-response brain activity, and cerebral blood flow. When samples of the different handwritings of a multiple are mixed with samples by different hands, police handwriting experts have been unable to identify them. There are data to suggest differences in allergic reactions and thyroid functioning. Drug studies have shown differences in responsiveness to alcohol and tranquillizers. Tests of memory have indicated genuine cross-personality amnesia for newly acquired information (while, interestingly enough, newly acquired motor skills are carried over).[12]

When and how did the multiplicity come into being?

The assumption made by most people in the MPD Movement—and which we so far have gone along with—is that the splitting into several selves (with all the sequelae we have been discussing) originates in early childhood.[13] The therapist therefore brings to light a pre-existing syndrome, and in no way is he (or she, for many therapists are women) responsible for creating MPD. But an alternative possibility of course exists, namely that the phenomenon—however genuine at the time that it is described—has been brought into being (and perhaps is being maintained) by the therapist himself.

We have hinted already at how little evidence there is that multiplicity has existed before the start of treatment. A lack of evidence that something exists is not evidence that it does not, and several papers at the Chicago meeting reported recently discovered cases of what seems to have been incipient multiplicity in children. Nonetheless, the suspicion must surely arise that MPD is an 'iatrogenic' condition (that is, generated by the doctor).

Folie à deux between doctor and patient would be, in the annals of psychiatry, nothing new.[14] It is now generally recognized that the outbreak of 'hysterical symptoms' in female patients at the end of the nineteenth century (including paralysis, anaesthesia, and so on) was brought about by the over-enthusiastic attention of doctors (such as Charcot) who succeeded in creating the symptoms they were looking for. In this regard, hypnosis, in particular, has always been a dangerous tool. The fact that in the diagnosis of multiplicity hypnosis is frequently (although not always) employed, the closeness of the therapist–patient relationship, and the intense interest shown by therapists in the 'drama' of MPD are clearly grounds for legitimate concern.

This concern is in fact one that senior members of the MPD Movement openly share. At the Chicago conference, a full day was given to discussing the problem of iatrogenesis. Speaker after speaker weighed in to warn their fellow therapists against 'fishing' for multiplicity, misuse of hypnosis, 'fascination' by the alter personalities, the 'Pygmalion effect', uncontrolled 'countertransference', and what was bravely called 'major league malpractice' (that is, sexual intimacy with patients). Although the message was that there is no need to invent the syndrome since you'll recognize the real thing when you see it, it is clear that those who have been in the business for some time understand only too well how easy it is to be misleading and misled.

A patient presents herself with a history of, let's call it, 'general muddle'. She is worried by odd juxtapositions and gaps in her life, by signs that she has sometimes behaved in ways that seem strange to her; she is worried she's going mad. Under hypnosis the therapist suggests that it is not her, but some other part of her that is the cause of trouble. And lo, some other part of her emerges. But since this is some other part, she requires—and hence acquires—another name. And since a person with a different name must be a different person, she requires—and hence acquires—another character. Easy; especially easy if the patient is the kind of person who is highly suggestible and readily dissociates, as is typical of those who have been subjected to abuse.

Could something like this possibly be the background to almost every case of MPD? We defer to the best and most experienced therapists in saying that it could not. In some cases there seems to be no question that the alternate personality makes its debut in therapy as if already formed. We have seen a videotape of one case where, in the first and only session of hypnosis, a pathetic young woman, Bonny, underwent a remarkable transformation into a character, calling herself

'Death', who shouted murderous threats against both Bonny and the hypnotist. Bonny had previously made frequent suicide attempts, of which she denied any knowledge. Bonny subsequently tried to kill another patient on the hospital ward and was discovered by a nurse lapping her victim's blood. It would be difficult to write off Bonny/Death as the invention of an overeager therapist.

On the general run of cases, we can only withhold judgement, not just because we do not know the facts, but also because we are not sure a 'judgemental' judgement is in order. Certainly we do not want to align ourselves with those who would jump to the conclusion that if MPD arises in the clinic rather than in a childhood situation it cannot be 'real'. The parallel with hysteria is worth pursuing. As Charcot himself demonstrated only too convincingly, a woman who feels no pain when a pin is stuck into her arm feels no pain—and calling her lack of reaction a 'hysterical symptom' does not make it any the less remarkable. Likewise a woman who at the age of thirty is now living the life of several different selves is now living the life of several different selves—and any doubts we might have about how she came to be that way should not blind us to the fact that such is now the way she is.

According to the model we proposed, no one starts off as either multiple or single. In every case there has to be some sort of external influence that tips the balance this way or that (or back again). Childhood may indeed be the most vulnerable phase; but it may also very well be that in certain people a state of incipient multiplicity persists much longer, not coming to fruition until later life.

The following story is instructive. A patient, Frances, who is now completely integrated, was telling us about the family of selves she used to live with—among whom she counted Rachel, Esther, Daniel, Sarah, and Rebecca. We were curious as to why a white Anglo-Saxon Protestant should have taken on these Hebrew names, and asked her where the names had come from. 'That's simple,' she said. 'Dad used to play Nazis and Jews with me; but he wanted me to be an innocent

victim, so every time he raped me he gave me a new Jewish
name.'

Here, it seems, that (as with Mary) the abuser at the time of
the abuse explicitly, even if unwittingly, suggested the personality structure of MPD. But suppose that Frances had not had
the 'help' of her father in reaching this 'solution'. Suppose she
had remained in a state of self-confusion, muddling through
her first thirty years until a sympathetic therapist provided her
with a way out (and a way forward). Would Frances have been
less of a multiple than she turned out to be? In our view, No.

There must be of course a world of difference between an
abuser's and a therapist's intentions in suggesting that a person contains several separate selves. Nonetheless, the consequences for the structure of the patient/victim's mind would
not be so dissimilar. 'Patrogenic' and 'iatrogenic' multiplicity
could be—and in our view would be—equally real.

Forty years ago, two early commentators, W. S. Taylor and
M. F. Martin, wrote:

Apparently most ready to accept multiple personality are (a) persons
who are very naive and (b) persons who have worked with cases or
near cases.[15]

The same is still largely true today. Indeed, the medical world
remains in general hostile to—even contemptuous of—MPD.
Why?

We have pointed to several of the reasons. The phenomenon is considered by many people to be scientifically or
philosophically absurd. We think that is a mistake. It is considered to be unsupported by objective evidence. We think
that is untrue. It is considered to be an iatrogenic folly. We
think that, even where that's so, the syndrome is a real one
nonetheless.

But there is another reason, which we cannot brush aside:
and that is the cliquish—almost cultish—character of those
who currently espouse the cause of MPD. In a world where
those who are not for MPD are against it, it is perhaps not sur-

prising that 'believers' have tended to close ranks. Maybe it is not surprising either that at meetings like the one we attended in Chicago there is a certain amount of well-meaning exaggeration and one-upmanship. We were, however, not prepared for what—if it occurred in a church—would amount to 'bearing witness'.

'How many multiples have you got?' one therapist asks another over breakfast in Chicago, 'I'm on my fifth.' 'Oh, I'm just a novice—two, so far.' 'You know Dr Q—she's got fifteen in treatment; and I gather she's a multiple herself.' At lunch: 'I've got a patient whose eyes change colour.' 'I've got one whose different personalities speak six different languages, none of which they could possibly have learned.' 'My patient Myra had her fallopian tubes tied, but when she switched to Katey she got pregnant.' At supper: 'Her parents got her to breed babies for human sacrifice; she was a surrogate mother three times before her eighteenth birthday.' 'At three years old, Peter was made to kill his baby brother and eat his flesh.' 'There's a lot of it about: they reckon that a quarter of our patients have been victims of satanic rituals.'

To be fair, this kind of gossip belies the deeper seriousness of the majority of therapists who deal with MPD. But that it occurs at all, and is seemingly so little challenged, could well explain why people outside the Movement want to keep their distance. Not to put too fine a point on it, there is everywhere the sense that both therapists and patients are participators in a Mystery to which ordinary standards of objectivity do not apply. Multiplicity is seen as a semi-inspired, semi-heroic condition: and almost every claim relating either to the patients' abilities or to the extent of their childhood suffering is listened to in sympathetic awe. Some therapists clearly consider it a privilege to be close to such extraordinary human beings (and the more of them in treatment, the more status the therapist acquires).

We were struck by the fact that some of the very specialists who have conducted the scientific investigations we mentioned earlier are sympathetic also to wild claims. We frankly cannot accept the truth of many of the circulating stories, and in particular we were unimpressed by this year's favourite, namely, all the talk of the 'satanic cult' origins of many cases of MPD.

However, an astronomer who believes in astrology would not for that reason be untrustworthy as an astronomical observer, and it would be wrong to find the phenomenon of multiplicity guilty by association. The climate in which the discussion is currently occurring is regrettable but probably unavoidable, not because all the true believers are gullible and all the opponents narrow-minded, but because those who have worked with cases know they have seen something so remarkable as to defy conventional description, and, in the absence of an accepted conceptual framework for description, they are driven by a sense of fidelity to their own experience to making hyperbolic claims.

We draw, for the time being, the following conclusions.

1. While the unitary solution to the problem of human selfhood is for most people socially and psychologically desirable, it may not always be attainable.

2. The possibility of developing multiple selves is inherent in every human being. Multiplicity is not only biologically and psychologically plausible, but in some cases it may be the best—even the only—available way of coping with a person's life experience.

3. Childhood trauma (usually, though not necessarily, sexual) is especially likely to push a person towards incipient multiplicity. It is possible that the child may progress from there to becoming a full-fledged multiple of his or her own accord; but in general it seems more likely that external pressure—or sanction—is required.

4. The diagnosis of MPD has become, within a particular psychiatric lobby, a diagnostic fad. Although the existence of the clinical syndrome is now beyond dis-

pute, there is as yet no certainty as to how much of the multiplicity currently being reported has existed prior to therapeutic intervention.

5. Whatever the particular history, the end result would appear to be in many cases a person who is genuinely split. That is, the grounds for assigning several selves to such a human being can be as good as—indeed, the same as—those for assigning a single self to a normal human being.

It remains the case that even in North America, the diagnosis of MPD has become common only recently, and elsewhere in the world it is still seldom made at all. We must surely assume that the predisposing factors have always been widely present in the human population. So where has all the multiplicity been hiding?

To end with further questions, and not answer them, may be the best way of conveying where we ourselves have got to. Here are some (almost random) puzzles that occur to us about the wider cultural significance of the phenomenon.

In many parts of the world the initiation of children into adult society has, in the past, involved cruel rites, including sexual and physical abuse (sodomy, mutilation, and other forms of battering). Is the effect (maybe even the intention) of such rites to create adults with a tendency to MPD? Are there contexts where an ability to split might be (or have been thought to be) a positive advantage—for example, when it comes to coping with physical or social hardship? Do multiples make better warriors?

In contemporary America, many hundreds of people claim to have been abducted by aliens from UFOs. The abduction experience is not recognized as such at first, and is described instead as 'missing time' for which the person has no memories. Under hypnosis, however, the subject typically recalls having been kidnapped by humanoid creatures who did harmful things to her or him—typically involving some kind of sex-related surgical operation (for example, sharp objects being thrust into the vagina). Are these people recounting a mythic version of an actual childhood experience? During the period described as missing time, was another personality in charge—a personality for whom the experience of abuse was all too real?

Plato banned actors from his Republic on the grounds that they were capable of 'transforming themselves into all sorts of characters'—a bad example, he thought, for solid citizens. Actors commonly talk about 'losing' themselves in their roles. How many of the best actors have been abused as children? For how many is acting a culturally sanctioned way of letting their multiplicity come out?

The therapists we talked to were struck by the 'charisma' of their patients. Charisma is often associated with a lack of personal boundaries, as if the subject is inviting everyone to share some part of him. How often have beguiling demagogues been multiples? Do we have here another explanation for the myth of 'the wound and the bow'?

Queen Elizabeth I, at the age of two, went through the experience of having her father, Henry VIII, cut off her mother's head. Elizabeth in later life was notoriously changeable, loving and vindictive. Was Elizabeth a multiple? Joan of Arc had trances, and cross-dressed as a boy. Was she?

5

Love Knots

Before St Valentine's day comes round again, it may be a good idea to brush up on a few facts about the saint. After all, you never know when the conversation over the candlelit dinner table may start to drag—and a flash of erudition may be just the ticket.

Now here's a fact. St Valentine, it seems, was not one saint, but two: one was a Roman priest, the other was bishop of Terni. They both lived in the fifth century AD, were both martyrs, died on the same day, and were buried in the same street of ancient Rome. Indeed, they were so much alike that no one can be sure they were not actually one and the same person— a 'doublet', as the *Dictionary of Saints* explains.[1]

'A "doublet", sweetheart?' 'Well, yes, that's presumably why it's *our* special day. Two people becoming one. Me and you united for ever. When you die, I die. Just like St Valentine and St Valentine.'

The bad news comes when I search further. 'There is,' the Dictionary says, 'nothing in either [sic] Valentine legend to account for the custom of choosing a partner of the opposite sex and sending "valentines" on 14 February.' Really? The *Encyclopaedia Britannica* confirms it: 'The association of the lovers' festival with St Valentine is purely accidental, and seems to arise from the fact that the feast of the saint falls in early spring.'[2]

Do they mean to tell us it could equally well have been St Colman's day (18 February) or St Polycarp's day (23 February)? Try sending your loved one a 'Colman'; try saying, 'Will you be my Polycarp?' Valentine, on the other hand, has the right music to it—it rhymes with thine, mine, and entwine. Valentine it clearly has to be.

But if the saint(s) were not responsible, who was? Encyclopaedias have their uses: 'VALENTINIAN I: Roman emperor . . . The great blot on his memory was his cruelty, which at times was frightful.' 'VALENTINIAN II: Son of the above . . . murdered in Gaul.' 'VALENTINIAN III: He was self-indulgent, incompetent, and vindictive.'

Oh dear. But who's this next? VALENTINUS and the VALENTIANS: 'Valentinus (2nd century) was the most prominent leader of the Gnostic movement . . . The lofty spirituality of the Gnostics degenerated over and over again into a distinctly sensual attitude. The chief sacrament of the Valentinians seems to have been that of the bridal chamber.'

Now, this is more like it. To become a Valentinian the initiate had to undergo a mystic marriage with the angel of death, the emissary of the great white mother goddess (who was represented as a sow). The priest said: 'Let the seed of light descend into thy bridal chamber; receive the bridegroom and give place to him, and open thine arms to embrace him . . . We must now become as one.'

So perhaps Valentine *is* all to do with two people merging into one. On which subject I was anyway intending to quote John Donne:

> Come forth, come forth, and as one glorious flame
>   Meeting another, grows the same,
>   So meet thy Fredericke, and so
> To an unseparable union grow.
>   . . . .
>
> And by this act of these two Phoenixes
>   Nature again restored is,
>   For since these two are two no more,
> There's but one Phoenix still, as was before.[3]

The poem was written to celebrate a Valentine's Day marriage. But was Donne in reality some kind of secret Valentinian Gnostic? He knew, it seems, about the death aspect of the marriage:

> A bride, before a good night could be said,
> Should vanish from her clothes, into her bed,
> As Souls from bodies steal.

He knew that love-making involved a symbolic resurrection:

> Up then fair Phoenix Bride, frustrate the Sun,
> . . . .
> Up, up, fair Bride, and call,
> Thy stars, from out their several boxes, take
> Thy rubies, pearls, and diamonds forth, and make
> Thy self a constellation, of them all,
> And by their blazing signify
> That a Great Princess falls, but doth not die.

Perhaps, too, he knew about the thirty concentric heavens (aeons) of Gnostic cosmology:

> But now she is laid; What though she be?
> Yet there are more delays, for, where is he?
> He comes, and passes through Sphere after Sphere.

Why 'Phoenixes'? Well, maybe because the mother goddess of the Gnostics originated in Phoenicia. But here is the real clincher. Valentinus, it turns out, was a candidate for the bishopric of Rome, but failed to be appointed. So Donne has wrapped it up in code with a play on the words 'not a Bishop' and 'Bishop's knot':

> You two have one way left, your selves to entwine,
> Besides this Bishop's knot, or Bishop Valentine.

Whether I shall get a footnote in the next collected works, I do not know. But I like to think that my honey-bunch will be suitably impressed.

6

Varieties of Altruism—and the Common Ground Between Them

Altruistic behaviour, where it occurs in nature, is commonly assumed to belong to one or other of two generically different types. Either it is an example of 'kin-selected altruism' such as occurs between blood relatives—a worker bee risking her life to help her sister, for example, or a human father giving protection to his child. Or it is an example of 'reciprocal altruism' such as occurs between non-relatives who have entered into a pact to exchange favours—one male monkey supporting another unrelated male in a fight over a female, for example, or one bat who has food to spare offering it to another unrelated individual who is hungry.

The first kind of altruism was given a theoretical explanation by William Hamilton, who showed how a gene that predisposes its carrier to help a close relative can prosper in the population, provided the genetic relationship between the two individuals is such that the cost to the giver is more than made up for by the benefit to the recipient multiplied by the degree of relatedness.[1] The second kind was given an explanation by Robert Trivers, who showed how a gene that predisposes its carrier to help a like-minded friend can also prosper, provided the social relationship between the individuals is such that the costs to the giver are more than made up for by the benefits that can be expected to be received later in exchange.[2]

These two kinds of textbook altruism—the free gift of help to a relative, or the exchange of help between friends—have for twenty years now been considered biologically, conceptually, and even morally distinct. In fact, so different are they supposed to be, that it has been claimed—on both sides—that only one of them should be counted genuinely 'altruistic'. Hamilton, for instance, in a recent review of the development of his own ideas has written:

My quest for biological altruism had carefully excluded anything I saw as possibly reciprocatory because it seemed that although behaviours of this category could mount a semblance of altruism, a donor always expected a benefit itself, at least in the long term: it was a semblance only. I still believe the reciprocal altruism that Trivers explained to me was misnamed.[3]

Trivers, by contrast, in the opening lines of his original paper on reciprocation defined altruistic behaviour as behaviour occurring specifically between individuals who are 'not closely related', and he played down altruistic acts within the family on the grounds that the altruist might 'merely be contributing to the survival of his own genes'. Models like Hamilton's, according to Trivers, were 'designed to take the altruism out of altruism'.[4]

Evidently the fathers of the two theories never doubted that the difference between them was a deep and important one. And, even with the moral question put aside, most later commentators have tended to agree. Although there is still some disagreement about the terms to use, almost everybody now accepts that there really are two very different things being talked about here: so that whenever we come across an example of helping behaviour in nature we can and ought to assign it firmly to one category or the other.

There have, however, been one or two dissenting voices. Stephen Rothstein, for example, in a little-known paper titled 'Reciprocal Altruism and Kin Selection Are Not Clearly Separable Phenomena', argued that—for reasons I shall develop further in a moment—many examples of reciprocal altruism must actually involve some degree of kin selection.[5] He was building on an earlier idea of Richard Dawkins's that two individuals who both carry copies of the same gene for altruism might be able to recognize one another by the very fact that they both tend to behave altruistically towards someone else—which in principle would allow them to promote their own genetic interests as altruists by selectively aiding each other (this being a special variant of what Dawkins called the 'green beard effect').[6] Dawkins wrote of this merely as a hypothetical possibility. But Rothstein realized that in many cases of reciprocal altruism it may in effect be close to the reality, since reciprocal altruists do in fact make a point of choosing other reciprocal altruists as trading partners.

Clearly, if Rothstein is right, reciprocal altruism may actually shade into kin selection. In which case the distinction between the two cannot be anything like so absolute as we have been led to believe. But suppose, now, there were to be further arguments in the same vein. Suppose it could be argued that kin selection also shades into reciprocal altruism. In that case we might want to challenge the reality of the distinction altogether, and might even wonder whether it would not be best to start the theoretical discussion over again. It is the purpose of the present essay to propose just such a radical rethink. For I believe Rothstein's paper did indeed tell only half the story, and that, when the other side is included too, it becomes clear that we can no longer continue looking at the landscape of altruism in the way we have grown used to. Rothstein asked us to recognize that altruistic behaviour towards friends must in many cases end up benefiting shared genes: but, I shall now argue, it is just as important we should recognize that altruistic behaviour towards kin must in many cases end up bringing a return of benefits to the altruist himself.

The arguments—Rothstein's and this new one of my own—are mirror images of each other, and, at the risk of being pedantic, I shall lay them out one after the other so as to show the structural similarities. But I shall begin with the new argument about how kin-selected altruism must often involve a degree of reciprocation, before giving my version of Rothstein's original argument about how reciprocal altruism involves a degree of kin selection.

Let us begin then by taking a new look at the case of kin-selected altruism: the case where we are dealing with an individual who has a gene that predisposes him to give help to a relative. Hamilton's famous point here was that every time this individual helps his relative he is benefiting any copy of the altruistic gene that the relative himself may happen to be carrying. So that, provided the cost, C, to the altruist is sufficiently small, and the benefit to the recipient, B, is sufficiently large, and there is in fact a sufficient degree of relationship between the two of them, r—provided, to be precise, that C < Br—the altruistic act will have provided a net gain in fitness to the gene. Which is why, in many circumstances, the gene is likely to evolve.

Hamilton's point is, of course, both valid and important. Nonetheless, I would now suggest that there has always been something nearly as important being ignored by any such simple analysis: namely, that every time the kin-selected altruist helps his relative he is also helping keep alive another individual, who, assuming he does have a copy of the altruistic gene, can be counted on to behave the same way towards *his* relatives—among whom is the original altruist himself. Hence the kin altruist, by helping his relative, is in effect increasing the pool of extant individuals from whom *he himself* may one day receive help.

Suppose, for example, that I have a gene that predisposes me to save my brother from drowning. Hamilton pointed out that in performing this altruistic act I have benefited any copy of my gene that is carried by my brother. But the new point is that I have in addition ensured the continued survival of someone who when the occasion arises is quite likely to save *me* from drowning. Therefore the pay-off to my fitness as a kin altruist will very likely come not only indirectly through the benefit to any copy of the altruism gene carried by my relative, but also directly through the benefit to my own gene.

It is true, as Hamilton would no doubt have wished to stress, that if I actually go so far as to commit suicide in order to save my brother, I will then miss out on any future return of benefits directly to myself. The same must in effect be true if I am already so old that I cannot expect to live for long or to have any more children in the future. In such cases, the return of benefits to me will only be able to come as a proxy benefit to one or other of my surviving descendants. Thus if I save a brother much younger than myself from drowning, it may be not me myself but, for example, my son (his nephew) who stands to benefit from his survival. But even here there is still going to be the chance of a significant return to my own direct line of genes.

The chances of either the altruist himself or any of his descendants getting this return of benefits is, of course, bound to depend on whether the relative who has been helped stays around long enough in the vicinity to be in a position to do his own bit of helping if and when the need arises. But this is likely to be much less of a problem than it might seem to be, since the very fact that the relative has received the earlier help is bound to increase his loyalty to the place and context in which it happened, and thus increase the chances that he will stay close to the altruist and/or his descendants. The fact that my brother, for example, has been saved by me from drowning is bound to encourage him to maintain close contact with me and my family in future years.

Let's be clear that it is not necessary to suppose that any kind of reciprocal-altruism-like 'bargain' is being struck in such cases. With these cases of kin altruism, the original helpful act can unquestionably be justified on Hamiltonian grounds alone, even if it never does bring any return to the altruist himself. The altruist certainly need not have any 'expectation' of getting anything in return, and the recipient need not feel under any 'moral compunction' to return the favour. Nonetheless, my point is that it will often so happen that the altruist *will* get the return.

Indeed, maybe it will so often so happen that a major part of the cost of the original altruistic act will as a matter of fact get repaid directly to the altruist. In which case it means that Hamilton's equation setting out the conditions under which this kind of altruism can be expected to evolve—his C < Br—has always been unduly pessimistic. For, in reality, the true net cost, C, of the original altruistic act will often work out in the long run to be much lower than at first it seems.

Now let us turn to the other side of the picture and take a closer look—as Rothstein did—at the case of the reciprocal altruist: the case where we are dealing with an individual who has a gene that predisposes him to help not a relative but rather a friend whom he trusts to pay him back. Trivers's famous point here was that every time this individual helps his friend he is adding to the stock of favours that are owed him. Hence, provided the cost of giving help to the friend in need is in general less than the benefit of receiving it when the altruist is in need himself, the exchange will have provided a net gain in fitness to the gene. Which is why, in many circumstances, this gene is likely to evolve.

Trivers's point, again, is true and important. But—and this was precisely Rothstein's argument—there has again been an important factor ignored by this analysis: namely, that every time the reciprocal altruist helps his friend he is also, so long as he has chosen wisely, increasing the chances of survival of another individual who is himself carrying the gene for reciprocal altruism. That is to say, he is increasing the chances of survival of another individual who, by carrying this gene, is in *this* respect a relative.

Suppose, for example, that I have a gene that predisposes me to save someone from drowning whom I think of as my friend. Trivers would say that in performing this altruistic act, I have behaved in a way likely to provide the friend with an incentive to return the favour to me in the future. But the new point is that I have also behaved in a way likely to bring immediate benefit to the gene that both my friend and I are carrying that has predisposed us to be friends to start with. Therefore the pay-off to my fitness as a reciprocal altruist is coming not only directly through the future benefit to my own gene, but also indirectly through the immediate benefit to his copy of it.

It is true, as Trivers himself would wish to stress, that if my friend were to happen to be a member of another species, a dog, say, rather than a human, this aspect of the pay-off would be wasted, since the dog's gene for reciprocal altruism is not part of my own species' gene pool and—even if it is functionally equivalent—it is presumably not transferable. Such cross-species friendships, in so far as they do occur in nature, must clearly be considered an exception to the point that is now being made. But this does nothing to weaken the argument as it relates to the much more common case of within-species friendships.

It needs to be said that, even with the within-species friendships, the copy of the gene for reciprocal altruism that can be assumed to be carried by each of the friends need not necessarily be the same gene by virtue of descent from a common ancestor—as it would be with true blood relatives. But there is no reason whatever why this should matter to natural selection. Indeed, for all that natural selection cares, one or other of the friends might actually be a first-generation reciprocal altruist who has acquired the gene by random mutation. All that matters is that the genes of the two friends have equivalent effects at the level of behaviour—and to suppose that only genes shared by common descent can count as being 'related' would, I think, be to fall into the conceit that philosophers have sometimes called 'origins chauvinism'.

Let's be clear again that it is not necessary to suppose that the probability of a kin- selection-like genetic pay-off in the case of friends has to be any part of their explicit motivation. The individual's act of reciprocal altruism can unquestionably be justified in the way that Trivers did originally, in terms of its expected return, without reference to any other possible effects on the fitness of the gene. Nonetheless, my point—and Rothstein's—is that in reality the indirect effect *will* often be there.

So much so that, again, as in the case of kin selection, it means that the standard model for how reciprocal altruism might evolve by natural selection may have seriously underestimated what there is going for it. In particular, the existence of an indirect benefit to the reciprocal altruism gene means that even if—because of bad luck or bad management—a particular altruistic act yields no return to the altruist, the effort put into it need still not have been entirely wasted. Trivers and his followers have tended to regard any such unrequited act of altruism as a disaster, and have therefore emphasized 'cheater-detection' as one of the primary concerns of social life. Yet the present analysis suggests that the system may in reality prove considerably more tolerant and more forgiving.

So, where does this leave us? We have clearly arrived at a rather different picture of the possibilities for altruism from the one that Hamilton and Trivers handed down. Instead of there being two fundamentally different types of altruistic behaviour, sustained by different forms of selection, we have discovered that each type typically has features of the other one. Kin altruism, even if primarily motivated by disinterested concern for the welfare of a relative, is often being selected partly because of the way it redounds to the altruist's own personal advantage. Reciprocal altruism, even if primarily motivated by the expectation of future personal reward, is often being selected partly because of the way it promotes the welfare of a gene-sharing friend.

In this case, there may be a further lesson to be learned. For if there is so much overlap between the two types of altruism, why should we continue to think in terms of *two* types at all? Might it not make more sense to suppose that at bottom all examples of altruism have a common formal structure—and a common basis at the level of the gene?

To be brief, I would suggest that the most revealing way of looking at the landscape of altruism is indeed to see it not as islands of kin altruism and reciprocal altruism, but as a continuum of possibilities that all have their roots in just one genetic trait: namely, a trait that is nothing else than *a trait for behaving altruistically to others who share this trait*. Or, to put this more expansively, *a trait for being helpful to those who can be expected to be helpful to those who can be expected to be helpful to those* . . . The recursiveness here is real and significant. It reflects precisely what happens when kin selection and reciprocation get combined. And it must add considerably to the chances of the trait becoming an evolutionary success.

The fact that all cases of altruism might be based on this one trait, however, should not lead us to expect that all cases should look alike in practice. For it is important to appreciate that the trait, as defined above, is only a semi-abstract formal disposition that still has to be realized at the level of behaviour. In particular, it still has to be decided how the possessor of the trait is going to be able to recognize who else counts as 'another individual who shares this trait'—who else counts, if you like, as 'one of us'—and hence who precisely should be the target of his or her own altruism.

The possibilities are various—and encompass both types of classical altruism we met with earlier. If, say, the target were to be identified solely on the basis of evidence of blood relationship ('There's a good chance she's one of us because she's my half-sister'), it would amount to an example of classical kin altruism. If, on the other hand, the target were to be identified solely on the basis of evidence of willingness to participate in friendly exchanges with oneself ('She's proved herself one of us by returning all the favours I've offered her'), it would amount to an example of classical reciprocal altruism.

But these would only be the two extremes, and in between would lie a range of other variations on the basic theme. There might be, for instance, a particular strain of altruists who identify their targets on the basis of evidence of altruistic behaviour directed to a third party ('She must be one of us because she's being so generous to them'). Another strain might identify them on the basis of the fact that they are already the targets of other altruists' behaviour ('She must be one of us because others of us are treating her as one of theirs'). And in a population where the altruistic trait has already evolved nearly to the point of fixation, it could even be that most altruists would identify their targets simply on the basis that they have not yet shown evidence of *not* being altruistically inclined ('Let's assume she's one of us until it turns out otherwise').

Not all these varieties of altruism would be evolutionarily stable under all conditions, and the two classical varieties probably do represent the two strategies that are evolutionarily safest. Nonetheless, others would prove adaptive, at least in the short term. And the best policy of all for any individual altruist would presumably be to mix and match different criteria for choosing targets, according to conditions.

We should therefore expect, in theory, to find altruism occurring in nature at many levels and in many different forms. This is a satisfying conclusion because—as must be obvious to anyone who can think in terms of more than the two original categories—in practice it is just what we do find. Frans de Waal in his compelling book, *Good Natured*, has detailed how wide-ranging and rich are the cooperative and succouring behaviours that are to be observed among non-human animals[7]—going far beyond what would seem to have been 'justified' by Hamilton's and Trivers's models. The same is more true still for human beings.

'The loveliest fairy in the world', Charles Kingsley wrote in *The Water-Babies*, 'is Mrs Doasyouwouldbedoneby.'[8] And she is also, as it happens, one of the most versatile and most successful.

FEELINGS

The Uses of Consciousness

In the picture (Fig. 3) is Denis Diderot—the eighteenth-century French philosopher, novelist, aesthetician, social historian, political theorist, and editor of the *Encyclopaedia*. It's hard to see how he had time, but alongside everything else, Diderot wrote a treatise called the *Elements of Physiology*—a patchwork of thoughts about animal and human nature, embryology, psychology and evolution. And tucked into this surprising work is this remark: 'If the union of a soul to a machine is impossible, let someone prove it to me. If it is possible, let someone tell me what would be the effects of this union.'[1]

Now, replace the word 'soul' with 'consciousness', and Diderot's two thought-questions become what are still the central issues in the science of mind. Could a *machine* be conscious? If it were conscious, what *difference* would it make?

The context for those questions is not hard to guess. Diderot was appalled by and simultaneously fascinated by the dualistic philosophy of René Descartes. 'A tolerably clever man', Diderot wrote,

began his book with these words: '*Man, like all animals, is composed of two distinct substances, the soul and the body.*' . . . I nearly shut the book. O! ridiculous writer, if I once admit these two distinct substances, you have nothing more to teach me. For you do not know what it is that you call soul, less still how they are united, nor how they act reciprocally on one another.[2]

Ridiculous it may have been. But fifty years later, the young Charles Darwin was still caught up with the idea: 'The soul,'

Fig. 3. Denis Diderot (1713–84)

he wrote in one of his early notebooks, 'by the consent of all is super-added.'[3]

This is one issue that the philosophy of mind has now done something to resolve. First has come the realization that there is no need to believe that consciousness is in fact something distinct from the activity of the physical brain. Rather, consciousness should be regarded as a 'surface feature' of the brain, an emergent property that arises out of the combined action of its parts. Second—and in some ways equally important—has come the realization that the human brain itself *is* a machine. So the question now is not *could* a machine be conscious or have a soul: clearly it could—I am such a machine, and so are you. Rather, the question is what *kind* of machine could be conscious. How much more and how much less would a conscious machine have to resemble the human brain—nerve cells, chemicals, and all? The dispute has become one between those who argue that it's simply a matter of having the appropriate 'computer programs', and those who say it's a matter of the 'hardware', too.

So-called 'functionalists', such as Daniel Dennett, argue that if a machine has whatever it takes at the level of the functional architecture for it to behave in all those ways that human beings do, then any such machine must by definition be conscious. But the 'non- functionalists' (or, as they are sometimes called, 'mysterians'), such as David Chalmers, argue that the mysterious *quality* of consciousness would still be missing.

This is an interesting dispute (see Chapter 9). And yet I'd say it clearly jumps the gun. It is all very well to discuss whether a machine which fulfils in every respect our *expectations* of how a conscious being *ought to behave* would actually be conscious. But the major question is still unresolved: what exactly *are* our expectations, and how might we account for them? In short, what do we think consciousness *produces*? If a machine could be united to a soul, what *effects*—if any— would it have?

When Diderot asked this crucial question, I think it is

obvious he was asking rhetorically for the answer 'None'. A machine, he was prepared to imagine, might have a soul—and yet for all practical purposes it would be indistinguishable from a machine without one: 'What difference,' he went on to ask, 'between a sensitive and living pocket watch and a watch of gold, of iron, of silver and of copper? If a soul were joined to the latter, what would it produce therein?'[4] Presumably, as a time-keeper—and that, after all, is what a watch does best— the watch would be just the same watch it was before: the soul would be no *use* to it, it wouldn't *show*.

I would not necessarily want to pin on to Diderot the authorship of the idea of the functional impotence of souls. But whenever it came, and whether or not Diderot got there, the realization that human consciousness itself might actually be useless was something of a breakthrough. I remember my own surprise and pleasure with this 'naughty' idea, when I first came across it in the writings of the behaviourist psychologists. There was J. B. Watson, in 1928, arguing that the science of psychology need make no reference to consciousness: 'The behaviourist sweeps aside all mediaeval conceptions. He drops from his scientific vocabulary all subjective terms such as sensation, perception, image, desire, and even thinking and emotion.'[5]

And there, as philosophical back-up, was Wittgenstein, arguing that concepts referring to internal states of mind have no place in the 'language game'.[6] If nothing else, it was an idea to tease one's schoolfriends with. 'How do I know that what I experience as the colour red, isn't what you experience as green? How do I know that you experience anything at all? You might be an unconscious zombie.'

A naughty idea is, however, all that it amounts to: an idea which has had a good run, and now can surely be dismissed. I shall give two reasons for dismissing it. One is a kind of Panglossian argument, to the effect that whatever exists as a consequence of evolution must have a function. The other is simply an appeal to common sense. But before I give either, let me say what I am *not* dismissing: I am not dismissing the idea that consciousness is a second-order and in some ways inessential process. In certain respects the behaviourists may have been right.

Diderot gives a nice example of *unconscious* behaviour:

A musician is at the clavecin; he is chatting with his neighbour, the conversation interests him, he forgets that he is playing a piece of concerted music with others; however, his eyes, his ear, his fingers are not the less in accord with them because of it; not a false note, not a misplaced harmony, not a rest forgotten, not the least fault in time, taste or measure. The conversation ceases, our musician returns to his part, loses his head and does not know where he has got to; the man is troubled, the animal is disconcerted. If the distraction of the man had continued for a few more minutes, the animal would have played the piece to the end, without the man having been aware of it.[7]

So the musician, if Diderot is right, sees without being aware of seeing, hears without being aware of hearing. Experimental psychologists have studied similar examples under controlled laboratory conditions and have confirmed that the phenomenon is just as Diderot described: while consciousness takes off in one direction, behaviour may sometimes go in quite another. Indeed, consciousness may be absent altogether. A sleep-walker, for example, may carry out elaborate actions and may even hold a simple conversation without waking up. Stranger things still can happen after brain injury. A person with damage to the visual cortex may lack all visual sensation, be consciously quite blind, and nonetheless be capable of 'guessing' what he would be seeing *if* he could see.[8] I have met such a case: a young man who maintained that he could see nothing at all to the left of his nose, and yet could drive a car through busy traffic without knowing how he did it.

So, that is what I am *not* dismissing: the possibility that the brain can carry on at least part of its job without consciousness being present. But what I *am* dismissing is the possibility that when consciousness *is* present it isn't making any difference. And let me now give the two reasons.

First, the evolutionary one. When Diderot posed his question, he knew nothing about Darwinian evolution.

He believed in evolution, all right—evolution of the most radical kind: 'The vegetable kingdom might well be and have been the first source of the animal kingdom, and have had its own source in the mineral kingdom; and the latter have originated from universal heterogeneous matter.'[9] What is more, Diderot had his own theory of selection, based on the idea of 'contradiction': 'Contradictory beings are those whose organization does not conform with the rest of the universe. Blind nature, which produces them, exterminates them; she lets only those exist which can co-exist tolerably with the general order.'[10]

Surprising stuff, seeing that it was written in the late eighteenth century. But note that, compared to the theory Darwin came up with eighty years later, there is something missing. Diderot's is a theory of *extinction*. According to him, the condition for a biological trait surviving is just that it should not contradict the general order, that it should not get in the way. Darwin's theory, on the other hand, is a theory of *adaptation*. According to him, the condition for something's surviving and spreading through the population is much stricter: it is not enough that the trait should simply be non-contradictory or neutral, it must—if it is to become in any way a general trait—be positively beneficial in promoting reproduction.

This may seem a small difference of emphasis, but it is crucial. For it means that when Diderot asked—of consciousness or anything else in nature—'What difference does it make?', he could reasonably answer: 'None'. But when a modern Darwinian biologist asks it, he cannot. The Darwinian's answer has to be that it has evolved because and only because it is serving some kind of useful biological function.

You may wonder, however: can we still expect consciousness to have a function even if we go along with the idea that it is in fact a 'mere surface feature' of the brain? But let's not be misled by the word 'mere'. We might say that the colours of a peacock's tail are a mere surface feature of the pigments, or that the insulating properties of fur are a mere surface feature of a hairy skin. But it is of course precisely on such surface features that natural selection acts: it is the colour or the warmth that matters to the animal's survival.

Philosophers have sometimes drawn a parallel between consciousness as a surface feature of the brain and wetness as a surface feature of water. Suppose we found an animal made entirely out of water. Its *wetness* would surely be the first thing for which an evolutionary biologist would seek to find a function.

Nonetheless, we do clearly have a problem: and this is to escape from a definition of consciousness that renders it self-evidently useless and irrelevant. Here philosophy of mind has, I think, been less than helpful. Too often we have been offered definitions of consciousness that effectively hamstring the enquiry before it has begun: for example, that consciousness consists in private states of mind of which the subject alone is aware, which can neither be confirmed nor contradicted, and so on. Wittgenstein's words, at the end of his *Tractatus*, have haunted philosophical discussion: 'Whereof one cannot speak, thereof one must be silent.'[11]

All I can say is that neither biologically nor psychologically does this feel right. Such definitions, at their limit (and they are meant of course to impose limits), would suggest that statements about consciousness can have no *information content*—technically, that they can do nothing to reduce anyone's uncertainty about what's going on. I find this counter-intuitive and wholly unconvincing. Which brings me to my second reason for dismissing the idea that consciousness is no use to human beings, which is that it is contrary to common sense.

Suppose I am a dentist, and I am uncertain whether the patient in the chair is feeling pain. I ask him, 'Does it hurt?', and he says, 'Yes. I'm not the kind of guy to show it, but it does *feel* awful'. Am I to believe that such an answer—as a description of a conscious state—contains no information? Common sense tells me that when a person describes his states of mind, either to me or to himself (not something he need be able to do, but something which as a matter of fact he often can do), he is making a revealing self-report. If he says, for example, 'I'm in pain', or 'I'm in love', or 'I'm having a green sensation', or 'I'm looking forward to my supper', I reckon I actually know more about him; but more important, that *through being conscious* he knows more about himself.

Still, the question remains: what sort of information is this? What is it about? And the difficulty seems to be that whatever it *is* about is, at least in the first place, private and subjective—something going on inside the subject which no one else can have direct access to. I think this difficulty has been greatly overplayed. There is, I'd suggest, an obvious answer to the question of what conscious descriptions are about: namely, that they are descriptions of what is happening inside the subject's *brain*. For sure, such information is 'private'. But it is private for the good reason that it happens to be his brain, hidden within his skull, and that he is naturally in a position to observe it in which the rest of us are not. Privacy is no doubt an issue of great biological and social significance, but I do not see that it is philosophically all that remarkable.

The suggestion that consciousness is a 'description of the brain' may nonetheless seem rather odd. Suppose someone says, for example, 'I'm not feeling myself today', that certainly doesn't sound like a description of a brain state. True enough, it does not *sound* like one; and no doubt I'd have trouble in persuading most people that it was so. Few people, if any, naturally make any connection between mind states and brain states. For one thing, almost no one except a brain scientist is likely to be interested in brains as such (and most people in the world probably don't even know they've got a brain). For another, there is clearly a huge gulf between brain states, as they are in fact described by brain scientists, and mind states as described by conscious human beings, a gulf which is practically—and, some would argue, logically—unbridgeable.

Yet is this really such a problem? Surely we are used to the idea that there can be completely different ways of describing the same thing. Light, for example, can be described either as particles *or* as waves, water can be described either as an aggregation of H_2O molecules *or* as a wet fluid, Ronald Reagan can be described either as an ageing movie actor *or* as the former President of the United States. The particular description we come up with depends on what measuring techniques we use and what our interests are. In that case, why should not the activity of the brain be described either as the electrical activity of nerve cells *or* as a conscious state of mind, depending on who is doing the describing? One thing is certain, and that is that brain scientists have different *techniques* and different *interests* from ordinary human beings.

I admit, however, I am guilty of some sleight of hand here. It is all very well to suggest that consciousness is 'a description' of the brain's activity by a subject with appropriate techniques and interests; but what I have not done is to locate this conscious subject anywhere. 'To describe' is a transitive verb. It requires a subject as well as an object, and they cannot in principle be one and the same entity. A brain, surely, cannot describe its own activity, any more than a bucket of water can describe itself as wet. In the case of the water, it takes an observer outside the bucket to recognize the water's wetness, and to do so he has to employ certain observational procedures—he has to stick his hand into it, swish it around, watch how it flows. Who, then, is the observer of the brain?

Oh dear. Are we stuck with an infinite regress? Do we need to postulate another brain to describe the first one, and then another brain to describe that? Diderot would have laughed:

If nature offers us a difficult knot to unravel, do not let us introduce in order to untie it the hand of a Being who then at once becomes an even more difficult knot to untie than the first one. Ask an Indian why the world stays suspended in space, and he will tell you that it is carried on the back of an elephant . . . and the elephant on a tortoise. And what supports the tortoise? . . . Confess your ignorance and spare me your elephant and your tortoise.[12]

You can hardly expect me, halfway through this essay, to

confess my ignorance. And in fact I shall do just the opposite. The problem of self-observation producing an infinite regress is, I think, phoney. No one would say that a person cannot use his own eyes to observe his own feet. No one would say, moreover, that he cannot use his own eyes, with the aid of a mirror, to observe his own eyes. Then why should anyone say a person cannot, at least in principle, use his own brain to observe his own brain? All that is required is that nature should have given him the equivalent of an *inner mirror* and an *inner eye*. And this, I think, is precisely what she has done. Nature has, in short, given to human beings the remarkable gift of *self-reflexive insight*. I propose to take this metaphor of 'insight' seriously. What is more, I even propose to draw a picture of it.

Imagine first the situation of an unconscious animal or a machine, which does not possess this faculty of insight (Figure 4). It has a brain which receives inputs from conventional sense organs and sends outputs to motor systems, and in between runs a highly sophisticated computer and decision maker. The animal may be highly intelligent and complexly motivated; it is by no means a purely reflex mechanism. But nonetheless it has no picture of what this brain-computer is doing or how it works. The animal is in effect an unconscious Cartesian automaton.

But now imagine (Figure 5) that a new form of sense organ evolves, an 'inner eye', whose field of view is not the outside

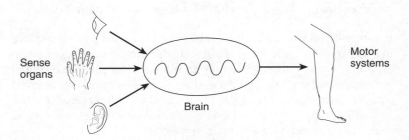

Fig. 4

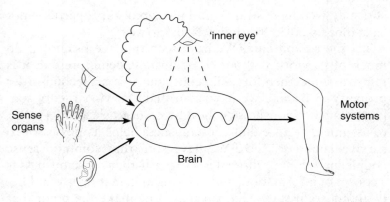

Fig. 5

world but the brain itself, as reflected via this loop. Like other sense organs, the inner eye provides a picture of its information field—the brain—which is partial and selective. But equally, like other sense organs, it has been designed by natural selection so that this picture is a useful one—in current jargon, a 'user-friendly' description, designed to tell the subject as much as he requires to know in a form that he is predisposed to understand. Thus it allows him, from a position of extraordinary privilege, to see his own brain states *as* conscious states of mind. Now every intelligent action is accompanied by the *awareness* of the thought processes involved, every perception by an accompanying sensation, every emotion by a conscious feeling.

Suppose this is what consciousness amounts to. I have written of consciousness as a surface feature of the brain and so I think it is, but you will see now that I am suggesting it is a very special sort of surface feature. For what consciousness actually is, is a feature not of the whole brain but of this added self-reflective loop. Why this particular arrangement should have what we might call the 'transcendent', 'other-worldly' qualities of consciousness I do not know. But note that I have allowed here for one curious feature: *the output of the inner eye is part of its own input.* A self-referential system of this sort may well have strange and paradoxical properties—not least that so-called 'truth functions' go awry.[13]

Let me recapitulate. We have seen that the brain can do much of its work without consciousness being present; it is fair to assume, therefore, that consciousness is a second-order property of brains. We have seen that Darwin's theory suggests that consciousness evolved by natural selection; it is fair to assume therefore that consciousness helps its possessor to survive and reproduce. We have seen that common sense coupled to a bit of self-analysis suggests that consciousness is a source of information, and that this information is very likely about brain states. So, if I may now make the point that immediately follows, it is fair to assume that access to this kind of second-order information about one's own brain states helps a person to survive and reproduce.

This looks like progress; and we can relax somewhat. In fact the heavier part of what I have to say is over. You ought, however, to be still feeling thoroughly dissatisfied; and if you are not, you must have missed the point of this whole essay. I set out to ask what difference consciousness makes, and have concluded that through providing insight into the workings of the brain it enhances the chances of biological survival. Fair enough. But the question of course is: how?

The problem is this. We have an idea of what consciousness is doing, namely, giving the subject a picture of his own brain activity, but we have no idea yet about what *biological good* this does him in the wider context of his daily life. It is rather as though we had discovered that fur keeps a rabbit warm, but had no idea of why a rabbit should *want* to keep warm. Or, to make a more relevant analogy, it is as though we had discovered that bats have an elaborate system for gathering information about echoes, but had no idea of why they should want such information.

The bat case provides a useful lesson. When Donald Griffin did his pioneering work on echo-location in bats, he did not of course first discover the echo-locating apparatus and then look for a function for it.[14] He began with the natural history of bats. He noted that bats live largely in the dark, and that their whole lifestyle depends on their apparently mysterious capacity to see without the use of eyes. Hence, when Griffin began his investigation of bats' ears and face and brain, he knew exactly what he was looking for: a mechanism within the bat which would allow it to 'listen in the dark'—and when he discovered such a mechanism there was of course no problem in deciding what its function was.

This is precisely the tactic we should adopt with consciousness in human beings. Having got this far, we should turn to natural history and ask: is there anything about the specifically human lifestyle which suggests that people, quite as much as bats, possess a mysterious capacity for understanding their natural environment, for which consciousness could be providing the mechanism?

I shall cut short a long story. When the question is, 'What would a natural historian notice as being special about the human life-style?', I'd say the answer must be this: human beings are extraordinarily *sociable* creatures. The environment to which they are adapted is before all else the environment of the family, the working group, the clan. Human interpersonal relationships have a depth, a complexity and a biological importance that far exceed those of any other animal. Indeed, without the ability *to understand, predict, and manipulate the behaviour* of other members of his own species, a person could hardly survive from day to day.

Now, this being so, it means that every individual has to be, in effect, a 'psychologist' just to stay alive, let alone to negotiate the maze of social interactions on which his success at mating and breeding will ultimately rest. Not a psychologist in the ordinary sense, but what I have called a 'natural psychologist'. Just as a blind bat develops quite naturally the ability to find its way around a cave, so every human being must develop a set of natural skills for penetrating the twilight world of interpersonal psychology—the world of loves, hates, jealousies, a world where so little is revealed on the surface and so much has to be surmised.

But this, when you think about, *is* rather mysterious. Because psychological understanding is immensely difficult; and understanding at the level that most people clearly have it would not, I suspect, be possible at all unless each individual had access to some kind of 'black-box' model of the human mind—a way of imagining what might be happening inside another person's head. In short, psychological understanding becomes possible because and only because people naturally conceive of other people as beings *with minds*. They attribute to them mental states—moods, thoughts, sensations, and so on—and it is on just this basis that they claim to understand them. 'She's *sad* because she *thinks* he doesn't *love* her', 'He's *angry* because he *suspects* she's *telling lies*', and so on across the range of human interaction.

I shall not, of course, pretend that this is news. If it were, it clearly would not be correct. But what we have to ask is where this ordinary, everyday, taken-for-granted psychological model of other human beings originates. How come that people latch on so quickly and apparently so effortlessly to seeing other people in this way? They do so, I suggest, because that is first of all *the way each individual sees himself*. And why is that first of all the way he sees himself? Because nature has given him an *inner eye*.

So here at last is a worthy function for self-reflexive insight. What consciousness does is to provide human beings with an extraordinarily effective tool for doing natural psychology. Each person can look in his own mind, observe and analyse his own past and present mental states, and on this basis make inspired guesses about the minds of others.

Try it . . . There is a painting by Ilya Repin that hangs in the Tretyakov Gallery in Moscow, its title *They did not expect him*. In slow motion, this is how I myself interpret the human content of the scene:

A man—still in his coat, dirty boots—enters a drawing room. The maid is apprehensive. She could close the door; but she doesn't—she wants to see how he's

Fig. 6. Ilya Repin, *They Did Not Expect Him* (1884).
Tretyakov Gallery, Moscow.

received. The grandmother stands, alarmed, as though
she's seen a ghost. The younger woman—eyes wide—
registers delighted disbelief. The girl—taking her cue
from the grown-ups—is suddenly shy. Only the boy
shows open pleasure. Who is he? Perhaps the father of
the family. They thought he'd been taken away. And now
he's walked in, as if from the dead. His mother can't
believe it; his wife didn't dare hope; the son was secretly
confident that he'd return. Where's he been? The maid's
face shows a degree of disapproval; the son's, excited
pride. The man's eyes, tired and staring, tell of a night-
mare from which he himself is only beginning to emerge.

The painting represents, as it happens, a Russian political prisoner, who has been released from the Tsar's jails and has come back home. We may not catch the final nuance—more information is needed. But try constructing or interpreting a scene like that *without* reference to consciousness, to what *we know* of human feelings—and the depth, its human depth, completely disappears.

I give this example to illustrate just how clever we all are. Consider those psychological concepts we've just 'called to mind'—apprehension, disbelief, disapproval, weariness, and so on. They are concepts of such subtlety that I doubt that any of us could explain in words just what they mean. Yet in dissecting this scene—or any other human situation—we wield them with remarkable authority. We do so because we have first experienced their meaning in ourselves.

It works. But I won't hide that there is a problem still of *why* it works. Perhaps we do, as I just said, wield these mental concepts 'with remarkable authority'. Yet who or what gives us this authority *to put ourselves in other people's shoes?* By what philosophical licence—if there is one—do we trespass so nonchalantly upon the territory of 'other minds'?

I am reminded of a story. There was dock strike in London, and enormous lorries were going in and out across the picket lines with impressive notices, 'By the Authority of H. M. Government', 'By the Permission of the Trades Union Congress', 'By the Authority of the Ministry of War'. Among them appeared a tiny donkey cart, driven by a little old man in a bashed-in bowler hat, and on the cart was the banner: 'By my own bloody authority'.[15]

That is a good plain answer to the problem. And yet I will not pretend that it will do. Tell a philosopher that ordinary people bridge this gap from self to other 'by their own bloody authority', and it will only confirm his worst suspicions that the whole business of natural psychology is flawed. Back will come Wittgenstein's objection that in the matter of mental states, one's own authority is no authority at all:

Suppose that everyone has a box with something in it; we call this thing a 'beetle'. No one can look into anyone else's box, and everyone says he knows what a beetle is only by looking at *his* beetle . . . [I]t would be quite possible for everyone to have something different in his box . . . [T]he box might even be empty.[16]

The problem, of course, is not entirely trivial. Strictly speaking, it is true we can never be sure that any of our guesses about the inner life of other people are correct. In a worst-case scenario, it is even possible that nature might have played a dreadful trick on us and built every human being according to a different plan. It is not just that the phenomenology of inner experience might differ from one person to another; the whole functional meaning of the experience might conceivably be different. Suppose, for example, that when *I* feel pain I do my best to stop it, but that when *you* feel pain you want more of it. In that case my own mental model—as a guide to your behaviour—would be useless.

This worst-case scenario is, however, one which as biologists we can totally discount. For the fact is—it is a biological fact, and philosophers ought sometimes to pay more attention than they do to biology—that human beings are all members of the same biological species: all descended within recent history from common stock, all still having more than 99.9 per cent of their genes in common, and all with brains which—at birth at least—could be interchanged without anyone being much the wiser. It is no more likely that two people will differ radically in the way their brains work than that they will differ radically in the way their kidneys work. Indeed, in one way it is—if I am right—even less likely. For while it is of no interest to a person to have the same kind of kidney as another person, it *is* of interest to him to have the same kind of mind: otherwise as a natural psychologist he would be in trouble. Kidney transplants occur very rarely in nature, but something very much like mind transplants occur all the time: you and I have just undergone one with those people in the painting. If the possibility of, shall we call it, 'radical mental polymorphism' had ever actually arisen in the course of human evolution, I think we can be sure that it would quickly have been quashed.

So that is the first and simplest reason why this method of doing psychology can work: the fact of the *structural similarity* of human brains. But it is not the only reason, nor in my view the most interesting one. Suppose that all human beings actually had identical brains, so that literally everything a particular individual could know about his own brain would be true of other people's: it could still be that his picture of his own brain would be no help in reading other people's behaviour. Why? Because it might just be the wrong kind of picture: it might be psychologically irrelevant. Suppose that when an individual looks in on his brain he were to discover that the mechanism for speech lies in his left hemisphere, or that his memories are stored as changes in RNA molecules, or that when he sees a red light there's a nerve cell that fires at 100 cycles per second. All of those things would very likely be true of other people too, but how much use would be *this* kind of inner picture as a basis for human understanding?

I want to go back for a moment to my diagram of the inner eye (Figure 5). When I described what I thought the inner eye does, I said that it provides a picture of its information field that has been designed by natural selection to be a useful one—a user-friendly description, designed to tell the subject as much as he requires to know. But at that stage I was vague about what exactly was implied by those crucial words, 'useful', 'user-friendly', 'requires to know'. I had to be vague, because the nature of the 'user' was still undefined and his specific requirements still unknown. By now, however, we have, I hope, moved on. Indeed, I'd suggest we now know exactly the nature of the user. The user of the inner eye is a natural psychologist. His requirement is that he should build up a model of the behaviour of other human beings.

This is where the natural selection of the inner eye has almost certainly been crucial. For we can assume that throughout a long history of evolution all sorts of different ways of describing the brain's activity have in fact been experimented with—including quite possibly a straightforward physiological description in terms of nerve cells, RNA, and so on. What has happened, however, is that only those descriptions most suited to doing psychology have been preserved. Thus the particular picture of our inner selves that human beings do in fact now have—the picture we know as 'us', and cannot imagine being of any different kind—is neither a *necessary* description nor *any old* description of the brain: it is the one that has proved most suited to our needs as social beings.

That is why it works. Not only can we count on other people's brains being very much like ours, we can count on the picture we each have of what it's like to have a brain being tailor-made to explain the way that other people actually behave. Consciousness is a socio-biological product—in the best sense of socio and biological.

So, at last, what difference does it make? It makes, I suspect, nothing less than the difference between being a man and being a monkey: the difference between we human beings *who know what it is like to be ourselves* and other creatures who essentially have no idea. 'One day,' Diderot wrote, 'it will be shown that consciousness is a characteristic of all beings.'[17] I am sorry to say I think that he was wrong. I recognize, of course, that human beings are not the only social animals on earth; and I recognize that there are many other animals that require at least a primitive ability to do psychology. But how many animals require anything like the level of psychological understanding that we humans have? How many can be said to require, as a biological necessity, a picture of what is happening inside their brains? And if they do not require it, why ever should they have it? What would a frog, or even a cow, lose if it were unable to look in on itself and observe its own mind at work?

I have, I should say, discussed this matter with my dog, and perhaps I can relay to you a version of how our conversation might have gone.

DOG. Nick, you and your friends seem to be awfully interested in this thing you call *consciousness*. You're always talking about it instead of going for walks.

NICK. Yes, well it is interesting, don't you think so?

DOG. You ask me that! You're not even sure I've got it.

NICK. That's why it's interesting.

DOG. Rabbits! Seriously, though, *do* you think I've got it? What could I do to convince you?

NICK. Try me.

DOG. Suppose I stood on my back legs, like a person? Would that convince you?

NICK. No.

DOG. Suppose I did something cleverer. Suppose I beat you at chess.

NICK. You might be a chess-playing computer. I'm very fond of you, but how do I know you're not just a furry soft automaton?

DOG. Don't get personal.

NICK. I'm not getting personal. Just the opposite, in fact.

DOG (*gloomily*). I don't know why I started this conversation. You're just trying to hurt my feelings.

NICK (*startled*). What's that you said?

DOG. Nothing. I'm just a soft automaton. It's all right for you. You don't have to go around *wishing* you were conscious. You don't have to feel *jealous* of other people all the time, in case they've got something that you haven't. And don't pretend you don't know what it feels like.

NICK. Yes, *I* know what it feels like. The question is, do *you*?

And this, I think, *remains* the question. I need hardly say that dogs, as a matter of fact, do not think (or talk) like this. Do any animals? Yes, there is some evidence that the great apes do: chimpanzees are capable of self-reference to their internal states, and can use what they know to interpret what others may be thinking.[18] Dogs, I suspect, are on the edge of it—although the evidence is not too good. But for the vast majority of other less socially sophisticated animals, not only is there no evidence that they have this kind of conscious insight, there is every reason to think that it would be a waste of time.

For human beings, however, so far from being a waste of time, it was the crucial adaptation—the sine qua non of their advancement to the human state. Imagine the biological benefits to the first of our ancestors who developed the capacity to read the minds of others by reading their own—to picture, as if from the inside, what other members of their social group were thinking about and planning to do next. The way was open to a new deal in social relationships, to sympathy, compassion, trust, deviousness, double-crossing, belief and disbelief in others' motives . . . the very things that make us human.

The way was open to something else that makes us human (and which my dog was quite right to pick up on): an abiding interest in the problem of what consciousness *is* and *why* we have it—sufficient, it seems, to drive biologically normal human beings to sit in a dim hall and listen to a lecture when they could otherwise have been walking in the park.

8

Farewell, Thou Art Too Dear for My Possessing

Farewell, thou art too dear for my possessing,
And like enough thou know'st thy estimate.
The charter of thy worth gives thee releasing;
My bonds in thee are all determinate.
For how do I hold thee but by thy granting,
And for that riches where is my deserving?
The cause of this fair gift in me is wanting,
And so my patent back again is swerving.
Thy self thou gav'st, thy own worth then not knowing,
Or me, to whom thou gav'st it, else mistaking;
So thy great gift, upon misprision growing,
Comes home again, on better judgement making.
   Thus have I had thee as a dream doth flatter:
   In sleep a king, but waking no such matter.

Shakespeare, *Sonnet LXXXVII*

I find this a disconcerting poem. Generous, tragic, but still a wet and slippery poem. Like an oyster, it slips down live—it's inside me, part of me, before I've had a chance to question it or chew it over. 'Farewell, thou art too dear for my possessing . . . Thus have I had thee as a dream doth flatter: / in sleep a king.' Exactly . . . Yes . . . But yes, exactly what? Why is the feeling in it so familiar? Who is the poem written to, what is it about, where does the feeling in it come from?

It would be a mistake, I think, to go for strictly factual answers—answers based on what we know or guess of Shakespeare's life. None of us, hearing the poem can hope to respond to it *as* Shakespeare: in the first instance, anyway, we respond to it as *us*. And our response to this as to any other work of art will not necessarily be any the less valid for our not knowing what the artist really meant. When I look at a seashell, mother-of-pearl glinting in the sun, I have no interest in the life story of the snail that made it. When I cup the seashell to my ear, it is not the snail I hear, it is not even the echo of the sea, it is simply the echo of myself.

To understand this poem, we do not *need* to know the facts of Shakespeare's life. But Shakespeare being Shakespeare, there is little chance we shall be *allowed* to remain ignorant for long. Suppose—just suppose—that this poem had really reached us from the blue, washed up on a beach in an old Elizabethan bottle. Suppose we knew nothing of its provenance, neither who wrote it, to whom it was written, or when. Before we could say Dr Rowse, a hundred Shakespeare scholars would have gathered round: a hundred detectives to scrape away the barnacles, each to emerge with his own—the only possible—historical interpretation. Shakespeare betrayed and abandoned by his friend. Shakespeare rejected by the patron who had supported him through the lean years of the plague. Marlowe, Shakespeare's rival, dead in a pub brawl.

Oh yes? It is not just that I do not feel the need to know. It is that actually I do not believe a word of it. Wrong poem, wrong bottle. Think about it. Just listen to what it is that Shakespeare is saying: 'For how do I hold thee but by thy granting, / And for that riches where is my deserving? / The cause of this fair gift in me is wanting.' Whatever else, it is surely not an honest description of Shakespeare's relation to his lover, let alone his relation to a rival playwright, or to whoever paid him for his poems. Conceivably Shakespeare's parting from a lover or a patron might have provided *the occasion* for the poem: such a parting might even—possibly—have provided a trigger for those feelings. But there is a world of difference between the trigger and the trap it springs. If Shakespeare meant this poem to be about his real feelings for his friend, then he was fooling someone.

But perhaps the person he was fooling was himself. 'The reason why it is so difficult for a poet not to tell lies is that, in poetry, all facts and all beliefs cease to be true or false and become interesting possibilities.'[1] The poet W. H. Auden may have been inclined to think there was something very special about poets—about himself and Shakespeare. But the fact is we all tell lies, we all live in a world of interesting possibilities: and never more than when we assess our own relationships to others.

We fall in love, we fall out of love, we think ourselves honoured by someone's friendship, we think ourselves betrayed by someone else's infidelity. But much of the time we pay scant attention to the realities of who we or they are. So we live out our loves, hopes, fears in fantasy, inventing the connections which we cannot or will not see clearly, looking to our own imagination for a possible label, a possible description of our feelings and our situation. And our heads being full of half-hoped-for, half-remembered possibilities, we give them, if nothing else, a poetical reality: airy nothings for which we find in *this* relationship a local habitation and a name. 'I love you.' No, it is the idea of you I love. 'I grieve for you.' No, it is my idea of you I grieve for. It is not my relationship to you I am describing, it is my relationship to the present embodiment of a host of earlier hopes and dreams.

So who, in imagination is Shakespeare saying farewell to? There is only one person I can think of to whom the sentiments might possibly apply. Not a lover, not a patron—or rather, *both* a lover *and* a patron and much more besides: the partner of everybody's first 'affair', the first to give herself to us and the first to take herself away. I am talking, you will guess, of each of our own *mother*s.

'For how do I hold thee *but by thy granting*, / And for that riches *where is my deserving*?' 'Thy self thou gav'st, *thy own worth then not knowing*, / Or me to whom thou gav'st it, else mistaking.' But 'my bonds in thee are all determinate'—our attachment could only last so long. And I, who—mother's darling—once thought himself a king, have now grown up to find myself deserted. 'No such matter' . . . No such *mater*.

It is a relationship—and a parting—we all know. One which we all, as poets, carry with us into adulthood, colouring every subsequent love and parting we encounter. No wonder we find the poem so slippery: mysterious yet familiar.

We throw it back into the sea. It washes in again on the next tide.

9

How to Solve the Mind–Body Problem

Two hundred and fifty years ago, Denis Diderot, commenting on what makes a great natural philosopher, wrote:

> They have watched the operations of nature so often and so closely that they are able to guess what course she is likely to take, and that with a fair degree of accuracy, even when they take it into their heads to provoke her with the most outlandish experiments. So that the most important service they can render to [others] . . . is to pass on to them that spirit of divination by means of which it is possible to *smell out*, so to speak, methods that are still to be discovered, new experiments, unknown results.[1]

Whether Diderot would have claimed such a faculty in his own case is not made clear. But I think there is no question we should claim it for him. For, again and again, Diderot made astonishingly prescient comments about the future course of natural science. Not least, this:

> Just as in mathematics, all the properties of a curve turn out upon examination to be all the same property, but seen from different aspects, so in nature, when experimental science is more advanced, we shall come to see that all phenomena, whether of weight, elasticity, attraction, magnetism or electricity, are all merely aspects of a single state.[2]

Admittedly the grand unifying theory that Diderot looked forward to has not yet been constructed. And contemporary physicists are still uncertain whether such a theory of *everything* is possible even in principle. But, within the narrower field that constitutes the study of *mind* and *brain*, cognitive scientists are increasingly confident of its being possible to have a unifying theory of these *two* things.

They—we—all assume that the human mind and brain are, as Diderot anticipated, aspects of a single state: a single state, in fact, of the material world, which could in principle be fully described in terms of its microphysical components. We assume that each and every instance of a human mental state is *identical* to a brain state, mental state, m = brain state, b, meaning that the mental state and the brain state pick out the same thing at this microphysical level. And usually we further assume that the nature of this identity is such that each type of mental state is multiply realizable, meaning that instances of this one type can be identical to instances of several different types of brain states that happen to be functionally equivalent.

What's more, we have reason to be confident that these assumptions are factually correct. For, as experimental science grows more advanced, we are indeed *coming to see* that mind and brain are merely aspects of a single state. In particular, brain-imaging studies, appearing almost daily in the scientific journals, demonstrate in ever more detail how specific kinds of mental activity (as reported by a mindful subject) are precisely correlated with specific patterns of brain activity (as recorded by external instruments). *This* bit of the brain lights up when a man is in pain, *this* when he conjures up a visual image, *this* when he tries to remember which day of the week it is, and so on.

No doubt many of us would say we have known all along that such correspondences must in principle exist; so that our faith in mind–brain identity hardly needs these technicolour demonstrations. Even so, it is, to say the least, both satisfying and reassuring to see the statistical facts of the identity being established, as it were, right before our eyes.

Yet it's one thing to see *that* mind and brain are aspects of a single state, but quite another to see *why* they are. It's one thing to be convinced by the statistics, but another to understand—as surely we all eventually want to—the causal or logical principles involved. Even while we have all the evidence required for *inductive generalization*, we may still have no basis for *deductive explanation*.

Let's suppose, by analogy, that we were to come to see, through a series of 'atmospheric-imaging' experiments, that whenever there is a visible shaft of lightning in the air there is a corresponding electrical discharge. We might soon be confident that the lightning and the electrical discharge are aspects of one and the same thing, and we should certainly be able to predict the occurrence of lightning whenever there is the electrical discharge. Even so, we might still have not a clue about what *makes* an electrical discharge manifest also as lightning.

Likewise, we might one day have collected so much detailed information about mind–brain correlations that we can predict which mental state will supervene on any specific brain state. Even so, we might still have no idea as to the reasons why this brain state yields this mental state, and hence no way of deducing one from the other a priori.

But with lightning there could be—and of course historically there was—a way to progress to the next stage. The physico-chemical causes that underlie the identity could be discovered through further experimental research and new theorizing. Now the question is whether the same strategy will work for mind and brain.

When experimental science is even more advanced, shall we come to see not only *that* mind and brain are merely aspects of a single state, but *why* they have to be so? Indeed, shall we be able to see how an identity that might otherwise appear to be mysteriously contingent is in fact transparently necessary?

A few philosophers believe the answer must be No. Or, at any rate, they believe we shall never achieve this level of understanding for every single feature of the mind and brain. They would point out that not all identities are in fact open to analysis in logical or causal terms, even in principle. Some identities are metaphysically primitive, and have simply to be taken as givens. And quite possibly some basic features of the mind are in this class. David Chalmers, for example, takes this stance when he argues for a version of epiphenomenal dualism in which consciousness just happens to be a fundamental, non-derivative property of matter.[3]

But even supposing—as most people do—that all the *interesting* identities are in fact analysable in principle, it might still be argued that not all of them will be open to analysis by us human beings. Thus Colin McGinn believes that the reason why a full understanding of the mind–brain identity will never be achieved is not because the task is logically impossible, but because there are certain kinds of understanding—and this is clearly one of them—which must for ever lie beyond our intellectual reach: no matter how much more factual knowledge we accumulate about mind and brain, we simply do not have what it would take to come up with the right theory.[4]

The poet Goethe, much earlier, counselled against what he considered to be the hubris of our believing that we humans can in fact solve every problem. 'In Nature,' he said,

there is an accessible element and an inaccessible . . . Anyone who does not appreciate this distinction may wrestle with the inaccessible for a lifetime without ever coming near to the truth. He who does recognize it and is sensible will keep to the accessible and by progress in every direction within a field and consolidation, may even be able to wrest something from the inaccessible along the way—though here he will in the end have to admit that some things can only be grasped up to a certain point, and that Nature always retains behind her something problematic which it is impossible to fathom with our inadequate human faculties.[5]

It is not yet clear how far—if at all—such warnings should be taken seriously. Diderot, for one, would have advised us to ignore them. Indeed Diderot, ever the scientific modernist, regarded any claim by philosophers to have found limits to our understanding, and thus to set up No-Go areas, as an invitation to science (or experimental philosophy) to prove such rationalist philosophy wrong.

Experimental philosophy knows neither what will come nor what will not come out of its labours; but it works on without relaxing. The philosophy based on reasoning, on the contrary, weighs possibilities, makes a pronouncement and stops short. It boldly said: 'light cannot be decomposed': experimental philosophy heard, and held its tongue in its presence for whole centuries; then suddenly it produced the prism, and said, 'light can be decomposed'.[6]

The hope now of cognitive scientists is, of course, that there is a prism awaiting discovery that will do for the mind–brain identity what Newton's prism did for light—a prism that will again send the philosophical doubters packing.

I am with them in this hope. But I am also very sure we shall be making a mistake if we ignore the philosophical warnings entirely. For there is no question that the likes of McGinn and Goethe might have a point. Indeed, I'd say they might have more than a point: they will actually become right by default, *unless and until we can set out the identity in a way that meets certain minimum standards for explanatory possibility.*

To be precise, we need to recognize that there can be no hope of scientific progress so long as we continue to write down the identity in such a way that the mind terms and the brain terms are patently *incommensurable.*[7] The problem will be especially obvious if the *dimensions* do not match up.

I use the word 'dimensions' here advisedly. When we do physics at school we are taught that the 'physical dimensions' of each side of an equation must be the same. If one side has the dimensions of a volume, the other side must be a volume, too, and it cannot be, for example, an acceleration; if one side has the dimensions of power, the other side must be power, too, and it cannot be momentum; and so on. As A. S. Ramsey put this in his classical *Dynamics* textbook: 'The consideration of dimensions is a useful check in dynamical work, for each side of an equation must represent the same physical thing and therefore must be of the same dimensions in mass [m], space [s] and time [t]'.[8]

Indeed, so strong a constraint is this that, as Ramsey went on, 'sometimes a consideration of dimensions alone is sufficient to determine the form of the answer to a problem'. For example, suppose we want to know the form of the equation that relates the energy contained in a lump of matter, E, to its mass, M, and the velocity of light, C. Since E can only have the dimension ms^2t^{-2}, M the dimension m and C the dimension st^{-1}, we can conclude without further ado that the equation must have the form $E = MC^2$. By the same token, if anyone were to propose instead that $E = MC^3$, we would know immediately that something was wrong.

But what is true of these dynamical equations is of course just as true of all other kinds of identity equations. We can be sure in advance that, if any proposed identity is to have even a chance of being valid, both sides must represent the same *kind* of thing. Indeed we can generalize this beyond physical dimensions to say that both sides must have the same conceptual dimensions, which is to say they must belong to the same generic class.

So, if it is suggested, for example, that Mark Twain and Samuel Clemens are identical, Mark Twain = Samuel Clemens, we can believe it because both sides of the equation are in fact people. Or, if it is suggested that Midsummer Day and 21 June are identical, Midsummer Day = 21 June, we can believe it because both sides are days of the year. But were someone to suggest that Mark Twain and Midsummer Day are identical, Mark Twain = Midsummer Day, we should know immediately this equation is a false one.

Now, to return to the mind–brain identity: when the proposal is that a certain mental state is identical to a certain brain state, mental state, m = brain state, b, the question is: do the dimensions of the two sides match?

The answer surely is, Yes, sometimes they do, or at any rate they can be made to.

Provided cognitive science delivers on its promise, it should soon be possible to characterize many mental states in computational or functional terms, that is, in terms of rules connecting inputs to outputs. But brain states too can relatively easily be described in these same terms. So it should then be quite straightforward, in principle, to get the two sides of the equation to line up.

Most of the states of interest to psychologists—states of remembering, perceiving, wanting, talking, thinking, and so on—are in fact likely to be amenable to this kind of functional analysis. So, although it is true there is still a long way to go before we can claim much success in practice, at least the research strategy is clear.

We do an experiment, say, in which we get subjects to recall what day of the week it is, and at the same time we record their brain activity by Magnetic Resonance Imaging (MRI). We discover that whenever a person thinks to himself, 'Today is Tuesday', a particular area of the brain lights up. We postulate the identity: recalling that today is Tuesday = activity of neurons in the calendula nucleus.

We then try, on the one hand, to provide a computational account of what is involved in this act of recalling the day; and, on the other hand, we examine the local brain activity and try to work out just what is being computed. Hopefully, when the results are in, it all matches nicely. A clear case of Mark Twain = Samuel Clemens.

But of course cases like this are notoriously the 'easy' cases—and they are not the ones that most philosophers are really fussed about. The 'hard' cases are precisely those where it seems that this kind of functional analysis is not likely to be possible. And this means especially those cases that involve *phenomenal consciousness*: the subjective sensation of redness, the taste of cheese, the pain of a headache, and so on. These are the mental states that Isaac Newton dubbed sensory 'phantasms',[9] and which are now more generally (although often less appropriately) spoken of as 'qualia'.

The difficulty in these latter cases is not that we cannot establish the factual evidence for the identity. Indeed, this part of the task may be just as easy as in the case of cognitive states such as remembering the day. We do an experiment, say, in which we get subjects to experience colour sensations, while again we examine their brain by MRI. We discover that whenever someone has a red sensation, there is activity in cortical area Q6. So we postulate the identity: phantasm of red = activity in Q6 cortex.

So far, so good. But it is the next step that is problematical. For now, if we try the same strategy as before and attempt to provide a functional description of the phantasm so as to be able to match it with a functional description of the brain state, the way is barred. No one, it seems, has the least idea how to characterize the phenomenal experience of redness in functional terms—or for that matter how to do it for any other variety of sensory phantasm. And in fact there are well-known arguments (such as the Inverted Spectrum) that purport to prove that it cannot be done, even in principle.

If not a functional description, then, might there be some other way of describing these elusive states which, being also applicable to brain states, could save the day? Unfortunately, the philosophical consensus seems to be that the answer must be No. For many philosophers seem to be persuaded that phenomenal states and brain states are indeed essentially such different kinds of entity that there is simply no room whatever for negotiation. Colin McGinn, in a fantasy dialogue, expressed the plain hopelessness of it sharply: 'Isn't it perfectly evident to you that . . . [the brain] is just the wrong kind of thing to give birth to [phenomenal] consciousness. You might as well assert that numbers emerge from biscuits or ethics from rhubarb.'[10] A case of Mark Twain = Midsummer Day.

Yet, as we've seen, this will not do! At least not if we are still looking for explanatory understanding. So, where are we scientists to turn?

Let's focus on the candidate identity: phantasm, p = brain state, b. Given that the statistical evidence supporting it remains as strong as ever, there would seem to be three ways that we can go.

1. We can accept that, despite everything, this equation is in fact false. Whatever the statistical evidence for there being a correlation between the two, there is really *not* an identity between the two states. Indeed, all the correlation shows is just that: that the states are *co-related*. And if we want to pursue it, we shall have then to go off and look for some other theoretical explanation for this correlation—God's whim, for instance. (This would have been Descartes's preferred solution.)

2. We can continue to believe in the equation, while at the same time grudgingly acknowledging that we have met our match: either the identity does not have an explanation or else the explanation really is beyond our human reach. And, recognizing now that there is no point in pursuing it, we shall be able, with good conscience, to retire and do something else. (This is McGinn's preferred solution.)

3. We can doggedly insist both that the identity is real and that we shall explain it somehow—when eventually we do find the way of bringing the dimensions into line. But then, despite the apparent barriers, we shall have to set to work to browbeat the terms on one side or other of the identity equation in such way as to *make* them line up. (This is my own and I hope a good many others' preferred solution.)

Now, if we do choose this third option, there are several possibilities.

One strategy would be to find a new way of conceiving of sensory phantasms so as to make them more obviously akin to brain states. But, let's be careful. We must not be *too* radical in redefining these phantasms or we shall be accused of redefining away the essential point. Daniel Dennett's sallies in this direction can be a warning to us.[11] His suggestion that sensations are nothing other than complex behavioural (even purely linguistic?) dispositions, while defensible in his own terms, has proved too far removed from most people's intuitions to be persuasive.

An alternative strategy would be to find a new way of conceiving of brain states so as to make them more like sensory phantasms. But again, we must not go too far. Roger Penrose is the offender this time.[12] His speculations about the brain as a quantum computer, however ingenious, have seemed to most neuroscientists to require too much special pleading to be taken seriously.

Or then again, there would be the option of doing *both*. My own view is that we should indeed try to meddle with both sides of the equation to bring them into line. Dennett expects all the compromise to come from the behavioural psychology of sensation, Penrose expects it all to come from the physics of brain states. Neither of these strategies seems likely to deliver what we want. But it's amazing how much more promising things look when we allow some give on *both* sides—when we attempt to adjust our concept of sensory phantasms *and* our concept of brain states until they do match up.

So *this*, I suppose, is how to solve the mind–brain identity problem. We shall need to work on both sides to define the relevant mental states and brain states in terms of concepts that really do have *dual currency*—being equally applicable to the mental and the material. And now all that remains, for this essay, is to do it.

Then let's begin: phantasm, p = brain state, b. Newton himself wrote: 'To determine . . . by what modes or actions light produceth in our minds the phantasms of colours is not so easy. And I shall not mingle conjectures with certainties.'[13] Three and a half centuries later, let us see if we can at least mix some certainties with the conjectures.

First, on one side of the equation, there are these sensory phantasms. Precisely what are we are talking about here? What kind of thing are they? What indeed are their dimensions?

Philosophers are—or at any rate have become in recent years—remarkably cavalier about the need for careful definition in this area. They bandy about terms such as 'phenomenal properties', 'what it's like', 'conscious feelings', and so on, to refer to whatever it is that is at issue when people point inwardly to their sensory experience—as if the hard-won lessons of positivist philosophy had never been learned. In particular, that overworked term 'qualia', which did at least once have the merit of meaning something precise (even if possibly vacuous),[14] is now widely used as a catch-all term for anything vaguely subjective and qualitative.

It is no wonder, then, that working scientists, having been abandoned by those who might have been their pilots, have tended to lose their way even more comprehensively. Francis Crick and Christoph Koch, for example, begin a recent paper by saying that 'everyone has a rough idea of what is meant by consciousness' and that 'it is better to avoid a precise definition of consciousness'.[15] In the same vein, Susan Greenfield writes, 'consciousness is impossible to define . . . [P]erhaps then it is simply best to give a hazy description, something like consciousness being "your first-person, personal world".'[16] Antonio Damasio is fuzzier still: 'Quite candidly, this first problem of consciousness is the problem of how we get a "movie in the brain" . . . [T]he fundamental components of the images in the movie metaphor are thus made of qualia.'[17]

But this is bad. Hazy or imprecise descriptions can only be a recipe for trouble. And, anyway, they are unnecessary. For the fact is we have for a long time had the conceptual tools for seeing through the haze and distinguishing the phenomenon of central interest.

Try this. Look at a red screen, and consider what mental states you are experiencing. Now let the screen suddenly turn blue, and notice how things change. The important point to note is that there are *two* quite distinct parts to the experience, and *two* things that change.

First (and I mean first), there is a change in the experience of something happening to yourself—the bodily sensation of the quality of light arriving at your eye. Second, there is a change in your attitude towards something in the outer world—your perception of the colour of an external object.

It was Thomas Reid, one of the geniuses of the Scottish enlightenment, who over two hundred years ago first drew philosophical attention to the remarkable fact that we human beings—and presumably many other animals also—do in fact use our senses in these two quite different ways:

The external senses have a double province—to make us feel, and to make us perceive. They furnish us with a variety of sensations, some pleasant, others painful, and others indifferent; at the same time they give us a conception and an invincible belief of the existence of external objects.

Sensation, taken by itself, implies neither the conception nor belief of any external object. It supposes a sentient being, and a certain manner in which that being is affected; but it supposes no more. Perception implies a conviction and belief of something external - something different both from the mind that perceives, and the act of perception. Things so different in their nature ought to be distinguished.[18]

For example, Reid said, we smell a rose, and two separate and parallel things happen: we both feel the sweet smell at our own nostrils and we perceive the external presence of a rose. Or, again, we hear a hooter blowing from the valley below: we both feel the booming sound at our own ears and we perceive the external presence of a ship down in the Firth. In general we can and usually do use the evidence of sensory stimulation *both* to provide a 'subject-centred affect-laden representation of what's happening to me', *and* to provide 'an objective, affectively neutral representation of what's happening out there'.[19]

Now it seems quite clear that what we are after when we try to distinguish and define the realm of sensory phantasms is the first of these: sensation rather than perception. Yet one reason why we find it so hard to do the job properly is that it is so easy to muddle the two up. Reid again:

[Yet] the perception and its corresponding sensation are produced at the same time. In our experience we never find them disjoined. Hence, we are led to consider them as one thing, to give them one name, and to confound their different attributes. It becomes very difficult to separate them in thought, to attend to each by itself, and to attribute nothing to it which belongs to the other. To do this, requires a degree of attention to what passes in our own minds, and a talent for distinguishing things that differ, which is not to be expected in the vulgar, and is even rarely found in philosophers.

I shall conclude this chapter by observing that, as the confounding our sensations with that perception of external objects which is constantly conjoined with them, has been the occasion of most of the errors and false theories of philosophers with regard to the senses; so the distinguishing these operations seems to me to be the key that leads to a right understanding of both.[20]

To repeat: sensation has to do with the self, with bodily stimulation, with feelings about what's happening *now* to *me* and how *I* feel about it; perception, by contrast, has to do with judgements about the objective facts of the external world. Things so different in their nature *ought* to be distinguished. Yet rarely are they. Indeed, many people still assume that perceptual judgements, and even beliefs, desires, and thoughts, can have a pseudo-sensory phenomenology in their own right.

Philosophers will be found claiming, for example, that 'there is something it is like' not only to have sensations such as feeling warmth on one's skin, but also to have perceptions such as seeing the shape of a distant cube, and even to hold propositional attitudes such as believing that Paris is the capital of France.[21] Meanwhile psychologists, adopting a half-understood vocabulary borrowed from philosophy, talk all too casually about such hybrid notions as the perception of 'dog qualia' on looking at a picture of a dog.[22] While these category mistakes persist we might as well give up.

So this must be the first step: we have to mark off the phenomenon that interests us—sensation—and get the boundary *in the right place*. But then the real work of analysis begins.

For we must home in on what *kind of thing* we are dealing with.

Look at the red screen. You feel the red sensation. You perceive the red screen. We do in fact talk of both sensation and perception in structurally similar ways. We talk of *feeling* or *having* sensations—as if somehow these sensations, like perceptions, were the *objects* of our sensing, sense *data*, out there waiting for us to grasp them or observe them with our mind's eye.

But, as Reid long ago recognized, our language misleads us here. In truth, sensations are no more the objects of sensing than, say, volitions are the objects of willing, intentions the objects of intending, or thoughts the object of thinking.

Thus, *I feel a pain*; *I see a tree*: the first denoteth a sensation, the last a perception. The grammatical analysis of both expressions is the same: for both consist of an active verb and an object. But, if we attend to the things signified by these expressions, we shall find that, in the first, the distinction between the act and the object is not real but grammatical; in the second, the distinction is not only grammatical but real.

The form of the expression, *I feel pain*, might seem to imply that the feeling is something distinct from the pain felt; yet in reality, there is no distinction. As *thinking a thought* is an expression which could signify no more than *thinking*, so *feeling a pain* signifies no more than *being pained*. What we have said of pain is applicable to every other mere sensation.[23]

So sensory awareness is an *activity*. We do not *have* pains, we *get to be* pained.

This is an extraordinarily sophisticated insight. And all the more remarkable that Reid should have come to it two hundred years before Wittgenstein was tearing his hair about similar problems and not getting noticeably further forward.

Even so, I believe Reid himself got only part way to the truth here. For my own view (developed in detail in my book, *A History of the Mind*)[24] is that the right expression is not so much 'being pained' as 'paining'. That is to say, sensing is not a passive state at all, but rather a form of active engagement with the stimulus occurring at the body surface.

When, for example, I feel pain in my hand, or taste salt on my tongue, or equally when I have a red sensation at my eye, I am not *being* pained, or *being* stimulated saltily, or *being* stimulated redly. In each case I am in fact the active agent. I am not sitting there passively absorbing what comes in *from* the body surface, I am reflexly reaching out *to* the body surface with an evaluative response—a response appropriate to the stimulus and the body part affected.

Furthermore, it is this *efferent activity* that I am aware of. So what I actually experience as the feeling—the sensation of what is happening to me—is my reading of my own response to it. Hence the quality of the experience, the way it feels, instead of revealing the way something is being done to me, reveals the very way something is being done by me.

This is how I feel about what's happening right now at my hand—I'm feeling painily about it!

This is how I feel about what's happening right now at this part of the field of my eye—I'm feeling redly about it!

In *A History of the Mind*, I proposed that we should call the activity of sensing 'sentition'. The term has not caught on. But I bring it up again here, in passing, because I believe we can well do with a word that captures the active nature of sensation: and sentition, resonating as it does with volition and cognition, sounds the right note of directed self-involvement.

The idea, to say it again, is that this sentition involves the subject 'reaching out to the body surface with an evaluative response—a response appropriate to the stimulus and the body part affected'. This should not of course be taken to imply that such sensory responses actually result in overt bodily behaviour—at least certainly not in human beings as we are now. Nonetheless, I think there is good reason to suppose that the responses we make today have in fact *evolved* from responses that in the past did carry through into actual behaviour. And the result is that even today the experience of sensation retains many of the original characteristics of the experience of true bodily action.

Let's consider, for example, the following five defining properties of the experience of sensation—and, in each case, let's compare an example of sensing, *feeling a pain in my hand*, with an example of bodily action, *performing a hand-wave*.

1. *Ownership.* Sensation always *belongs to the subject.* When I have the pain my hand, I *own* the paining, it's mine and no one else's, I am the one and only *author* of it; as when I wave my hand, I *own and am the author of* the action of waving.
2. *Bodily location.* Sensation is always *indexical* and *invokes a particular part of the subject's body.* When I have the pain in my hand, the paining intrinsically involves *this* part of *me*; as when I wave my hand, the waving, too, intrinsically involves *this* part of *me.*
3. *Presentness.* Sensation is always *present tense, ongoing, and imperfect.* When I have the pain in my hand, the paining is in existence just *now for the time being*; as when I wave my hand, the waving, too, exists just *now.*
4. *Qualitative modality.* Sensation always has the feel of one of several *qualitatively distinct modalities.* When I have the pain in my hand, the paining belongs to the class of *somatic* sensations, quite different in their whole style from, say, the class of visual sensations or of olfactory ones; as when I wave my hand, the waving belongs to the class of *hand-waves*, quite different in style from other classes of bodily actions such as, say, the class of face-smiles or of knee-jerks.
5. *Phenomenal immediacy.* Most important, sensation is always *phenomenally immediate*, and the four properties above are *self-disclosing.* Thus, when I have the pain in my hand, my impression is simply that *my hand hurts*: and, when my hand hurts, the fact that it is *my* hand (rather than someone else's), that it is my *hand* (rather than some other bit of me), that it is hurting *now* (rather than some other time), and that it is

acting in a *painful* fashion (rather than acting in a visual, gustatory, or auditory fashion) are facts of which I am directly and immediately aware *for the very reason that it is I, the author of the paining, who make these facts*; just as when I wave my hand, my impression is simply that my hand waves, and all the corresponding properties of this action too are facts of which I, *the author of the wave*, am immediately aware for similar reasons.

Thus, in these ways and others that I could point to, the positive analogies between sensations and bodily activities add up. And yet, I acknowledge right away that there is also an obvious disanalogy: namely that, to revert to that old phrase, it is 'like something' to have sensations, but not like anything much to engage in most other bodily activities.

To say the least, our experience of other bodily activities is usually very much shallower. When I wave my hand, there may be, perhaps, the ghost of some phenomenal experience. But surely what it's like to wave hardly compares with what it's like to feel pain, or taste salt, or sense red. The bodily activity comes across as a flat and papery phenomenon, whereas the sensation seems so much more velvety and thick. The bodily activity is like an unvoiced whisper, whereas the sensation is like the rich, *self-confirming* sound of a piano with the sustaining pedal down.

Of course, neither metaphor quite captures the difference in quality I am alluding to. But still I think the sustaining pedal brings us surprisingly close. For I believe that ultimately the key to an experience being 'like something' does in fact lie in the experience *being like itself in time*—hence *being about itself*, or *taking itself as its own intentional object*. And this is achieved, in the special case of sensory responses, through a kind of *self-resonance* that effectively stretches out the present moment to create what I have called the *thick moment of consciousness*.[25]

There are, of course, loose ends to this analysis, and ambiguities. But I'd say there are surely fewer of both than we began with. And this is the time to take stock, and move on.

The task was to recast the terms on each side of the mind–brain identity equation, phantasm, p = brain state, b, so as to make them look more like each other.

What we have done so far is to redescribe the left-hand side of the equation in progressively more concrete terms. Thus the phantasm of pain becomes the sensation of pain, the sensation of pain becomes the experience of actively paining, the activity of paining becomes the activity of reaching out to the body surface in a painy way, and this activity becomes self-resonant and thick. And with each step we have surely come a little closer to specifying something of a *kind* that we can get a handle on.

We can therefore turn our attention to the right hand side of the equation. As Ramsey wrote, 'sometimes a consideration of dimensions alone is sufficient to determine the form of the answer to a problem'. If we now have this kind of thing on the mind side, we need to discover something like it on the brain side. If the mind term involves a state of *actively doing something about something*, namely issuing commands for an evaluative response addressed to body surface, then the brain term must also be a state of actively doing something about something, presumably doing the corresponding thing. If the mind term involves *self-resonance,* then the brain state must also involve self-resonance. And so on.

Is this still the impossibly tall order that it seemed to be earlier—still a case of ethics on one side, rhubarb on the other? No, I submit that the hard problem has in fact been transformed into a relatively easy problem. For we are now dealing with something on the mind side that surely *could* have the same dimensions as a brain state *could*. Concepts such as 'indexicality', 'present-tenseness', 'modal quality', and 'authorship' are indeed dual currency concepts of just the kind required.

It looks surprisingly good. We can surely now imagine what it would take on the brain side to make the identity work. But

I think there is double cause to be optimistic. For, as it turns out, this picture of what is needed on the brain side ties in beautifully with a plausible account of the evolution of sensations.

I shall round off this essay by sketching in this evolutionary history. And if I do it in what amounts to cartoon form, I trust this will at least be sufficient to let the major themes come through.

Let's return, then, in imagination to the earliest of times and imagine a primitive amoeba-like animal floating in the ancient seas.

This animal has a defining edge to it, a structural boundary. This boundary is crucial: the animal *exists* within this boundary—everything within it is part of the animal, belongs to it, is part of 'self', everything outside it is part of 'other'. The boundary holds the animal's own substance in and the rest of the world out. The boundary is the vital frontier across which exchanges of material and energy and information can take place.

Now light falls on the animal, objects bump into it, pressure waves press against it, chemicals stick to it. No doubt some of these surface events are going to be a good thing for the animal, others bad. If it is to survive, it must evolve the ability to sort out the good from the bad and to respond differently to them—reacting to this stimulus with an 'Ow!', to that with an 'Ouch!', to this with a 'Whowee!'.

Thus, when, say, salt arrives at its skin, it detects it and makes a characteristic wriggle of activity—it wriggles saltily. When red light falls on it, it makes a different kind of wriggle—it wriggles redly. These are adaptive responses, selected because they are appropriate to the animal's particular needs. Wriggling saltily has been selected as the best response to salt, while wriggling sugarly, for example, would be the best response to sugar. Wriggling redly has been selected as the best response to red light, while wriggling bluely would be the best response to blue light.

To begin with, these wriggles are entirely local responses, organized immediately around the site of stimulation. But later there develops something more like a reflex arc passing via a central ganglion or proto-brain: information arrives from the skin, it gets assessed, and appropriate adaptive action is taken.

Still, as yet, these sensory responses are nothing other than responses, and there is no reason to suppose that the animal is in any way mentally aware of what is happening. Let's imagine, however, that, as this animal's life becomes more complex, the time comes when it will indeed be advantageous for it to have some kind of inner knowledge of what is affecting it, which it can begin to use as a basis for more sophisticated planning and decision making. So it needs the capacity to form *mental representations* of the sensory stimulation at the surface of its body and how it feels about it.

Now, one way of developing this capacity might be to start over again with a completely fresh analysis of the incoming information from the sense organs. But this would be to miss a trick. For, the fact is that all the requisite details about the stimulation—where the stimulus is occurring, what kind of stimulus it is, and how it should be dealt with—are already encoded in the command signals the animal is issuing when it makes the appropriate sensory response.

Hence, all the animal needs to do to represent the stimulation is to pick up on these already-occurring command signals. For example, to sense the presence of salt at a certain location on its skin, it need only monitor its own signals for wriggling saltily at that location, or, equally, to sense the presence of red light it need only monitor its signals for wriggling redly.

Note well, however, that all this time the animal's concern is merely with what's occurring at its body surface. By monitoring its own responses, it forms a representation of 'what is happening to me'. But, at this stage, the animal neither knows nor cares *where the stimulation comes from*, let alone what the stimulation may imply about the world *beyond* its body.

Yet wouldn't it be better off if it *were* to care about the world beyond? Let's say a pressure wave presses against its side: wouldn't it be better off if, besides being aware of feeling the pressure wave as such, it were able to interpret this stimulus as signalling an approaching predator? A chemical odour drifts across its skin: wouldn't it be better off if it were able to interpret this stimulus as signalling the presence of a tasty worm? In short, wouldn't the animal be better off if, as well as reading the stimulation at its body surface merely in terms of its immediate affective value, it were able to interpret it as a *sign* of 'what is happening out there'?

The answer of course is, Yes. And we can be sure that, early on, animals did in fact hit on the idea of using the information contained in body surface stimulation for this novel purpose—*perception* in addition to *sensation*. But the purpose was indeed *so* novel that it meant a very different style of information-processing was needed. When the question is 'What is happening to me?', the answer that is wanted is qualitative, present-tense, transient, and subjective. When the question is 'What is happening out there?', the answer that is wanted is quantitative, analytical, permanent, and objective.

So, to cut a long story short, there developed in consequence two parallel channels to subserve the very different readings we now make of an event at the surface of the body, sensation and perception: one providing an affect-laden modality-specific body-centred representation of what the stimulation is doing to me and how I feel about it, the other providing a more neutral, abstract, body-independent representation of the outside world.

Sensation and perception continued along relatively independent paths in evolution. But we need not be concerned further with perception in this essay. For it is the fate of sensation that matters to our narrative.

As we left it, the animal is actively responding to stimulation with public bodily activity, and its experience or proto-experience of sensation (if we can now call it that) arises from its monitoring its own command signals for these sensory responses. Significantly, these responses are still tied in to the animal's survival and their form is still being maintained by natural selection—and it follows that the form of the animal's sensory experience is also at this stage being determined in all its aspects by selection.

Yet, the story is by no means over. For, as this animal continues to evolve and to change its lifestyle, the nature of the selection pressures is bound to alter. In particular, as the animal becomes more independent of its immediate environment, it has less and less to gain from the responses it has always been making directly to the surface stimulus as such. In fact, there comes a time when, for example, wriggling saltily or redly at the point of stimulation no longer has any adaptive value at all.

Then why not simply give up on this primitive kind of local responding altogether? The reason why not is that, even though the animal may no longer want to respond directly to the stimulation at its body surface as such, it still wants to be able to keep up to date mentally with what's occurring (not least because this level of sensory representation retains a crucial role in policing perception; see Chapter 10). So, even though the animal may no longer have any use for the sensory responses in themselves, it has by this time become quite dependent on the secondary representational functions that these responses have acquired. And since the way it has been getting these representations in the past has been by monitoring its own command signals for sensory responses, it clearly cannot afford to stop issuing these command signals entirely.

So, the situation now is this. In order to be able to represent 'what's happening to me', the animal must in fact continue to issue commands such as *would* produce an appropriate response at the right place on the body *if* they were to carry through into bodily behaviour. But, given that the behaviour is no longer wanted, it may be better if these commands remain virtual or as-if commands—in other words, commands which, while retaining their original intentional properties, do not in fact have any real effects.

The upshot is—or so I've argued—that, over evolutionary time, there is a slow but remarkable change. What happens is that the whole sensory activity gets 'privatized': the command signals for sensory responses get short-circuited before they reach the body surface, so that instead of reaching all the way out to the site of stimulation, they now reach only to points closer and closer in on the incoming sensory nerve, until eventually the whole process becomes closed off from the outside world in an internal loop within the brain (Fig. 7).

Now once *this* happens, the role of natural selection must of course sharply diminish. The sensory responses have lost all their original biological importance and have in fact disappeared from view. Therefore selection is no longer involved in determining the form of these responses and a fortiori it can no longer be involved in determining the quality of the representations based on them.

But the fact is that this privacy has come about only at the very end, after natural selection has done its work to shape the sensory landscape. There is therefore every reason to suppose that the forms of sensory responses and the corresponding experiences have already been more or less permanently fixed. And although, once selection becomes irrelevant, these forms may be liable to drift somewhat, they are likely always to reflect their evolutionary pedigree. Thus responses that start-

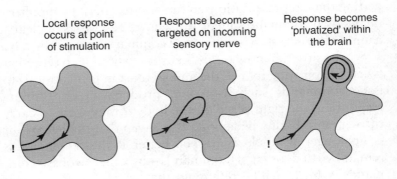

Local response occurs at point of stimulation

Response becomes targeted on incoming sensory nerve

Response becomes 'privatized' within the brain

Fig. 7

ed their evolutionary life as dedicated wriggles of acceptance or rejection of a stimulus will still be recognizably of their kind right down to the present day.

Yet, something is not in place yet: the 'thickness factor'. And, as it happens, there is a further remarkable evolutionary development to come—made possible by the progressive shortening of the sensory response pathway.

It has been true all along, ever since the days when sensory responses were indeed actual wriggles at the body surface, that they have been having *feedback* effects by modifying the very stimulation to which they are a response. In the early days, however, this feedback circuit was too roundabout and slow to have had any interesting consequences. However, as and when the process becomes internalized and the circuit so much shortened, the conditions are there for a significant degree of recursive interaction to come into play. That's to say, the command signals for sensory responses begin to loop back upon themselves, becoming in the process partly self-creating and self-sustaining. These signals still *take their cue* from input from the body surface, and still get *styled* by it, but on another level they have become signals *about themselves*. To be the author of such recursive signals is to enter a new intentional domain.

To return to our identity equation: we needed a certain set of features on the brain side. We could have invented them if we were brave enough. But now, I submit, we actually have them handed to us on a plate by an evolutionary story that delivers on every important point.

I acknowledge that there is more to be done. And the final solution to the mind–body problem, if ever we do agree on it, may still look rather different from the way I'm telling it here. But the fact remains that this approach to the problem has to be the right one. There is no escaping the need for dual currency concepts – and any future theory will have to play by these rules.

Diderot wrote:

A tolerably clever man began his book with these words: '*Man, like all animals, is composed of two distinct substances, the soul and the body. If anyone denies this proposition it is not for him that I write.*' I nearly shut the book. O! ridiculous writer, if I once admit these two distinct substances, you have nothing more to teach me.[26]

This essay has been about how to make one thing of these two.

10

The Privatization of Sensation

D. H. Lawrence, the novelist, once remarked that if anyone presumes to ask *why* the midday sky is blue rather than red, we should not even attempt to give a scientific answer but should simply reply: 'Because it *is*.' And if anyone were to go still further, and to ask why his own conscious sensation when he looks at the midday sky is characterized by blue qualia rather than red qualia, I've no doubt that Lawrence, if he were still around—along with several contemporary philosophers of mind—would be just as adamant that the last place we should look for enlightenment is science.

But this is not my view. The poet and critic William Empson wrote: 'Critics are of two sorts: those who merely relieve themselves against the flower of beauty, and those, less continent, who afterwards scratch it up. I myself, I must confess, aspire to the second of these classes; unexplained beauty arouses an irritation in me.'[1] And equally, I'd say, unexplained *subjective experience* arouses an irritation in *me*. It is the irritation of someone who is an unabashed Darwinian: one who holds that the theory of evolution by natural selection has given us the licence to ask 'why' questions about almost every aspect of the design of living nature, and, what's more, to expect that these 'whys' will nearly always translate into scientifically accredited 'wherefores'.

Our default assumption, I believe, can and should be that living things are designed the way they are because this kind of design is—or has been in the past—biologically advantageous. And this will be so across the whole of nature, even when we come to ask deep questions about the way the human mind works, and even when what's at issue are the central facts of consciousness.

Why is it like *this* to have red light fall on our eyes? Why like *this* to have a salt taste in our mouths? Why like *this* to hear a trumpet sounding in our ears? I think these questions, as much as any, deserve our best attempt to provide Darwinian answers: answers, that is, in terms of the biological function that is being—or has been—served.

There are two levels at which the questions can be put. First, we should ask about the biological function of our having *sensations at all*. And, next, once we have an answer to this first question, we can proceed to the trickier question about the function of our sensations being of *the special qualitative character they are*.

No doubt the first will strike most people as the easy question, and only the second as the hard one. But I admit that even this first question may not be as easy as it seems. And, although I want to spend most of this essay discussing sensory quality, I realize I ought to begin at the beginning by asking: what do we gain, of biological importance, from having sensations at all?

To see why this seemingly easy question requires serious consideration and why the answer is not in fact self-evident, we have to take on board the elementary distinction between sensation and perception.

The remarkable fact that human beings—and presumably many other animals also—make use of their bodily senses in two quite different ways was first brought to philosophical attention two hundred years ago by Thomas Reid. 'The external senses,' Reid wrote, 'have a double province—to make us feel, and to make us perceive.'[2] Which is to say (as elaborated in Chapter 9) we can and usually do use the evidence of sensory stimulation *both* to provide 'a subject-centred affect-laden representation of what's happening to me', *and* to provide 'an objective, affectively neutral representation of what's happening out there'.

Yet, while Reid insisted so firmly on this difference, he never, it seems, thought it necessary to ask the question that so clearly follows on: *why* do the senses have a double province? Do human beings really need both perception *and* sensation? If, as might well be argued—especially in the case of vision and hearing—what interests us in terms of our survival is not at all our personal relation to the stimulation at our body surface but only what this stimulation denotes about the outside world, why ever should we bother to represent 'what is happening to me' as well as 'what is happening out there'? Why should we not leave sensation out of it entirely and make do with perception on its own? Would not such insensate perception serve our biological needs perfectly well?

It is only in the last few years that psychologists have begun to face up to the genuine challenge of this question 'Why sensations?'. And there is certainly no agreement yet on what the right Darwinian answer is. However, there are now at least several possible answers in the offing. And I, Anthony Marcel, and Richard Gregory have all, in different ways, endorsed what is probably the strongest of these: namely, that sensations are required, in Gregory's felicitous wording, 'to flag the present'.[3]

The idea here is that the main role of sensations is, in effect, to help keep perception honest. Both sensation and perception take sensory stimulation as their starting point: yet, while sensation then proceeds to represent the stimulation more or less as given, perception takes off in a much more complex and risky way. Perception has to combine the evidence of stimulation with contextual information, memory, and rules so as to construct a hypothetical model of the external world as it exists independently of the observer. Yet the danger is that, if this kind of construction is allowed simply to run free, without being continually tied into present-tense reality, the perceiver may become lost in a world of hypotheticals and counterfactuals.

What the perceiver needs is the capacity to run some kind of on-line reality check, testing his perceptual model for its currency and relevance, and in particular keeping tabs on where he himself now stands. And this, so the argument goes, is precisely where low-level, unprocessed sensation does in fact prove its value. As I summarized it earlier: 'Sensation lends a here-ness and a now-ness and a me-ness to the experience of the world, of which pure perception in the absence of sensation is bereft.'[4]

I think we should be reasonably happy with this answer. The need to flag the present provides at least one compelling reason why natural selection should have chosen sensate human beings over insensate ones.

But we should be under no illusion about how far this answer takes us with the larger project. For it must be obvious that even if it can explain why sensations exist at all, it goes no way to explaining why sensations exist in the particular qualitative form they do.

The difficulty is this. Suppose sensations have indeed evolved to flag the present. Then surely it hardly matters precisely *how* they flag the present. Nothing would seem to dictate that, for example, the sensation by which each of us represents the presence of red light at our eye must have the particular red quality it actually does have. Surely this function could have been performed equally well by a sensation of green quality or some other quality completely?

Indeed, would not the same be true of any other functional role we attribute to sensations? For the fact is—isn't it?—that sensory quality is something private and ineffable, maybe of deep significance to each of us subjectively but of no consequence whatever to our standing in the outside world.

Certainly, there is a long philosophical tradition that makes exactly this claim. John Locke originated it with his thought experiment about the undetectable 'inverted spectrum'.[5] Imagine, said Locke, that 'if by the different structure of our organs, it were so ordered, that the same object should produce in several men's minds different ideas at the same time; e.g. if the idea, that a violet produces in one man's mind by his eyes, were the same that a marigold produced in another man's, and *vice versa*.' Then, Locke surmised, there's no reason to think this difference in inner structure and the resulting difference in the inner experience of the quality of colour would make any difference to outer behaviour. In fact, he claimed, the difference in inner experience 'could never be known: because one man's mind could not pass into another man's body'.

Ludwig Wittgenstein would later remark: 'The assumption would thus be possible—though unverifiable—that one section of mankind has one sensation of red and another section another.'[6] Indeed, this unsettling possibility became one of the chief reasons why Wittgenstein himself decided to call a halt to any further talk about privately sensed qualities. And it is the reason, too, why other philosophers such as Daniel Dennett have been tempted to go even further, and to argue that sensory qualia have no objective reality whatever.[7]

Now, we need not go all the way with Wittgenstein or Dennett to realize that if even part of this argument about the privacy of qualia goes through, we may as well give up on our ambition to have a Darwinian explanation of them. For it must be obvious that nothing can possibly have evolved by natural selection unless it does in fact have some sort of major public effect—indeed, unless it has a measurably positive influence on survival and reproduction. If, as common sense, let alone philosophy, suggests, sensory quality really is for all practical purposes private, selection simply could never have got a purchase on it.

It appears that we cannot have it both ways: *either* as Darwinists we continue, against the odds, to try to explain sensory quality as a product of selection, *or* we grudgingly accept the idea that sensations are just as private as they seem to be.

So, what is to be done? Which of these two strongly motivated positions must we give up?

I believe the answer is that actually we need not give up either. We can in fact hold *both* to the idea that sensory quality is private, *and* to the idea that it has been shaped by selection, provided we recognize that these two things *have not been true at the same time*: that, in the course of evolution, the privacy came only *after* the selection had occurred.

Here, in short, is the case that I would make. It may be true that the activity of sensing is today largely hidden from public view, and that the particular quality of sensations is not essential to the function they perform. It may be true, for example, that my sensation of red is directly known only to me, and that its particular redness is irrelevant to how it does its job. Yet, *it was not always so*. In the evolutionary past the activity of sensing was a much more open one, and its every aspect mattered to survival. In the past my ancestors evolved to feel red this way because feeling it this way gave them a real biological advantage.

Now, in case this sounds like a highly peculiar way of looking at history, I should stress that it would not be so unusual for evolution to have worked like this. Again and again in other areas of biology it turns out that, as the function of an organ or behaviour has shifted over evolutionary time, obsolete aspects of the original design have in fact carried on down more or less unchanged.

For a simple example, consider the composition of our own blood. When our fish ancestors were evolving four hundred million years ago in the Devonian seas, it was essential that the salt composition of their blood should closely resemble the external sea water, so that they would not lose water by osmosis across their gills. Once our ancestors moved on to the land, however, and started breathing air, this particular feature of blood was no longer of critical importance. Nevertheless, since other aspects of vertebrate physiology had developed to fit in with it and any change would have been at least temporarily disadvantageous, well was left alone. The result is that human blood is still today more or less interchangeable with sea water.

This tendency towards what can be called 'stylistic inertia' is evident at every level of evolution, not only in nature but in culture, too. Clear examples occur in the development of language, manners, dress, and architectural design (as has been beautifully documented by Philip Steadman).[8] But I would say that as nice a case as any is provided by the history of clocks and how their hands move.

Modern clocks have of course evolved from sundials. And in the Northern hemisphere, where clocks began, the shadow of the sundial's vane moves round the dial in the 'sunwise' direction which we now call 'clockwise'. Once sundials came to be replaced by clockwork mechanisms with moving hands, however, the reason for representing time by sunwise motion immediately vanished. Nevertheless, since by this stage people's habits of time-telling were already thoroughly ingrained, the result has been that nearly every clock on earth still does use sunwise motion.

But suppose now, for the sake of argument, we were to be faced with a modern clock, and, as inveterate Darwinians, we were to want to know *why* its hands move the way they do. As with sensations, there would be two levels at which the question could be posed.

If we were to ask about *why the clock has hands at all*, the answer would be relatively easy. Obviously the clock needs to have hands of some kind so as to have some way of representing the passage of time—just as we need to have sensations of some kind so as to have some way of representing stimulation at the body surface.

But if we ask about *why the hands move clockwise as they do*, the answer would have to go much deeper. For clearly the job of representing time could in fact nowadays be served equally well by rotationally inverted movement—just as the job of representing sensory stimulation could nowadays be served equally well by quality-inverted sensations. In fact, as we've seen, this second question for the clock can *only* be answered by reference to ancestral history—just as I would argue for sensations.

When an analogy fits the case as well as this, I would say it cries out to be taken further. For it strongly suggests there is some profounder basis for the resemblance than at first appears. And, in this instance, I believe we really have struck gold. The clock analogy provides the very key to what sensations are and how they have evolved.

A clock tells time by *acting* in a certain way, namely by moving its hands; and this action has a certain style inherited from the past, a clockwise style, clockwisely. The remarkable truth is, I believe, that a person also has sensations by *acting* in a certain way; and, yes, each sensory action also has its own inherited style—for example, a red style, redly.

This theory of 'sensations as actions'—bodily actions that in the course of evolution have become 'privatized'—is the subject of my book *A History of the Mind*, and is outlined in Chapter 9 (see especially pages 108–13). I shall summarize here only the crucial details.

Sensations, I argue, originate with the protective and appetitive responses of primitive animals to stimulation at their body surface. Under the influence of natural selection, these sensory responses become finely adapted so as to reflect the particular pattern of stimulation, with the intentional properties of the response varying according to the nature of the stimulus, its bodily location, its modality, and so on. These responses therefore *encode* information about what is happening at the body surface and how the animal evaluates it. However, in the early stages, the animal has no interest in *bringing this information to mind*—no use, as yet, for any kind of representation of 'what's happening to me'.

Life grows more complicated, and the animal, besides merely responding to stimulation, does develop an interest in forming *mental representations* of what's happening to it. But, given that the information is already encoded in its own sensory responses, the obvious way of getting there is to monitor its command signals for making these responses. In other words, the animal can, as it were, tune into 'what's happening to me and how I feel about it' by the simple trick of noting 'what I'm doing about it'.

The result is that sensations do indeed evolve at first as corollaries of the animal's *public* bodily activity. And since, in these early days, the form of this activity is still being maintained by natural selection, it follows that the forms of the animal's mental representations—its sensory 'experience' or proto-experience, if you like—are also being determined in all their aspects by selection.

Life moves on further, and the making of responses directly to stimuli at the body surface becomes of less and less relevance to the organism's biological survival. In fact, the time comes when it may be better to suppress these responses altogether. But now the animal is in something of a bind. For, while it may no longer have any need for the bodily responses as such, it has become quite dependent on the mental representations of the stimuli—representations that have been based on monitoring these very responses.

The animal has therefore to find a way of continuing to make *as if* to respond to the stimuli, without actually carrying the activity through into overt behaviour. And the upshot is— or so I've argued—that the whole sensory activity gets *privatized*, with the command signals getting short-circuited before they reach the body surface, and eventually getting entirely closed off from the outside world in an internal loop within the brain (Fig. 7 at Chapter 9, p. 112).

Once *this* happens, the role of natural selection must sharply diminish. The sensory responses have lost all their original biological importance and have in fact disappeared from view. Note well, however, that this privatization has come about only at the very end, *after* natural selection has done its work to shape the sensory landscape. And the forms of sensory responses and the corresponding experiences have already been more or less permanently fixed, so as to reflect their pedigree.

It is this evolutionary pedigree that still colours private sensory experience right down to the present day. If *I* today feel the sensation red *this* way—as *I* know very well that I do—it is because I am descended from distant ancestors who were selected to feel it this same way long ago.

Here we are, then, with the solution that I promised. We *can* have it both ways. We can *both* make good on our ambition to explain sensory quality as a product of selection, *and* we can accept the common-sense idea that sensations are as private as they seem to be—provided we do indeed recognize that these two things have not been true at the same time.

But the rewards of this Darwinian approach are, I believe, greater still. For there remains to be told the story of how, after the privatized command signals have begun to loop back on themselves within the brain, there are likely to be dramatic consequences for sensory phenomenology. In particular, how the activity of sensing is destined to become self-sustaining and partly self-creating, so that sensory experiences get lifted into a time dimension of their own—into what I have called the 'thick time' of the subjective present.[9] What is more, how the establishment of this time loop is the key to the thing we value most about sensations: the fact that not only do they *have* quality but that this quality comes across to us in the very special, self-intimating way that we call the *what it's like of consciousness*.

When did this transformation finally occur? Euan Macphail, among others, has argued that conscious sensations require the prior existence of a *self*.[10] The philosopher Gottlob Frege made a similar claim: 'An experience is impossible without an experient. The inner world presupposes the person whose inner world it is.'[11] I agree with both these writers about the requirement that sensations have a self to whom they belong. But I think Macphail, in particular, goes much too far with his insistence that such a self can only emerge with language. My own view is that self-representations arise through action, and that the 'feeling self' may actually be created by those very sensory activities that make up its experience.

This is, however, another story for another time. I will simply remark here, with Rudyard Kipling, *contra* Lawrence, that 'Them that asks no questions isn't told a lie'[12]—and no truths either.

DISCOVERIES

11

Mind in Nature

Outside the London Zoo the other day I saw a hydrogen-filled dolphin caught in a tree. It was bobbing about, blown by the wind, every so often making a little progress upwards, but with no prospect of escape. To the child who'd released it, I was tempted to explain: yes, the balloon would 'like' to rise into the air, but unfortunately it doesn't have much common sense.

Now imagine a real dolphin caught under a tree. Like the balloon, it might push and struggle for a time. But dolphins are clever. And having seen how matters stood, a real dolphin would work out a solution *in its mind*: soon enough it would dive and resurface somewhere else.

Dolphins have intelligence, balloons do not. A hundred years ago William James defined intelligence: 'With intelligent agents, altering the conditions changes the activity displayed, but not the end reached . . . The pursuance of future ends and *the choice of means for their attainment* are thus the mark and criterion of mentality in a phenomenon.'[1]

James was not interested in animals alone. He wrote of intelligent 'agents' and of mentality in a 'phenomenon'—and allowed that intelligence might belong to entities quite other than creatures like dolphins or ourselves. Since his time, we have indeed become used to the idea of intelligent man-made machines. But for James, as he progressed towards a mystical old age, the question became more open still. Does intelligence characterize the whole of Nature? Is the sea intelligent, is the weather, are galaxies . . . is God?

These issues have become hot topics again among philosopher-biologists, but none hotter than the question of the 'species mind'. Since Darwin's time, it has been realized that evolving species behave as if they have a goal—the goal of self-advancement in the struggle for survival. But until recently it's been assumed that the forces of evolution are no more intelligent or better informed than the balloon. Species push up against the environmental limits, but can go no further until some accidental change allows them to inch forward. What's more, evolving species have no foresight—they cannot see the dead end or the sharp twig (and indeed, like the balloon, they sometimes pop).

Now, however, new ideas are emerging with an emphasis on the ways a species may in fact play an 'intelligent' part in its own evolution. It is even being suggested that there is a sense in which a species may be able to 'think things through' before it ratifies a particular evolutionary advance.

The philosopher Jonathan Schull argues it like this.[2] The mark of higher intelligence, he notes, is that the solution to a problem comes in stages: (i) possible solutions are generated at random, (ii) those solutions likely to succeed are followed through in the imagination, (iii) only then is the best of these adopted in practice. But doesn't something very like this occur in biological evolution, too? It does, and it has a name: the 'Baldwin effect'.

Let's suppose the members of a species are confronted by a problem—the problem, say, of how to keep warm. By chance there arises a particular individual who learns by his own efforts how to build a cosy nest. Natural selection then gets to work and favours him and his successors who show equivalent inventiveness in nest-building. But, as a result of this, the process of invention begins to get short-circuited, because there is selection for anything that makes the invention more likely to occur. So more and more of the elements of *learned* nest-building (selecting the material, weaving it together, and so on) become incorporated in *genetic* programs—until eventually the learned habit has become an unlearned instinct. In Schull's words:

The adapting population of animals thus plays a role in evolution analogous to that of the cortex in individual mammalian behaviour, interacting with the environment to develop new patterns and try them out, and then passing control of routinized patterns to more automatic centres.

For those who want to see it, the parallel to intelligence is striking. Where we go with it is another matter. Does it mean that species, considered as entities in their own right, are in some sense consciously aware? Are individual organisms the species-equivalent of thoughts?

Several mystical traditions have claimed as much. For myself, I'm chary of any talk of species consciousness. But I like the idea of individuals as trial runs for the population, whose passing strokes of ingenuity have slowly become cast in DNA. It pleases me, too, to think that individual creativity may have lain behind much of the inherited design in nature. Perhaps as a species we owe more than we realize to those long-dead pioneers who once thought up the answers we're now born with: those individual heroes who, like Shakespeare's Mark Antony, 'dolphin-like showed their backs above the element they lived in'.[3]

12

Cave Art, Autism, and the Evolution of the Human Mind

'Man is a great miracle', the art historian Ernst Gombrich was moved to say, when writing about the newly discovered paintings at the Chauvet and Cosquer caves.[1] The paintings of Chauvet, especially, dating to about 30,000 years ago, have prompted many people to marvel at this early flowering of the modern human mind. Here, it has seemed, is clear evidence of a new kind of mind at work: a mind that, after so long a childhood in the Old Stone Age, had grown up as the mature, cognitively fluid mind we know today.

In particular it has been claimed that these and other examples of Ice Age art demonstrate, first, that their makers must have possessed high level conceptual thought: for example,

The Chauvet cave is testimony that modern humans . . . were capable of the type of symbolic thought and sophisticated visual representation that was beyond Neanderthals,[2]

or

Each of these painted animals . . . is the embodiment and essence of the animal species. The individual bison, for example, is a spiritual-psychic symbol; he is in a sense the 'father of the bison,' the idea of the bison, the 'bison as such'.[3]

Second, that their makers must have had a specific intention to represent and communicate information: for example,

The first cave paintings . . . are the first irrefutable expressions of a symbolic process that is capable of conveying a rich cultural heritage of images and probably stories from generation to generation,[4]

or, more particularly,

> This clearly deliberate and planned imagery functions to stress one part of the body, or the animal's activity . . . since it is these that are of interest [to the hunter].[5]

And, third, that there must have been a long tradition of artistry behind them: for example,

> We now know that more than 30,000 years ago ice age artists had acquired a complete mastery of their technical means, presumably based on a tradition extending much further into the past.[6]

The paintings and engravings must surely strike anyone as wondrous. Still, I draw attention here to evidence that suggests that the miracle they represent may not be at all of the kind most people think. Indeed this evidence suggests the very opposite: that the makers of these works of art may actually have had distinctly pre-modern minds, have been little given to symbolic thought, have had no great interest in communication, and have been essentially self-taught and untrained. Cave art, so far from being the sign of a new order of mentality, may perhaps better be thought the swan-song of the old.

The evidence I refer to, which has been available for more than twenty years now (although apparently unnoticed in this context) comes from a study made in the early 1970s by Lorna Selfe of the artwork of a young autistic girl named Nadia.[7]

Nadia, born in Nottingham in 1967, was in several respects severely retarded. By the age of six years she had still failed to develop any spoken language, was socially unresponsive, and was physically clumsy. But already in her third year she had begun to show an extraordinary drawing ability, suddenly starting to produce line drawings of animals and people, mostly from memory, with quite uncanny photographic accuracy and graphic fluency:

> Nadia's ability, apart from its being so superior to other children, was also essentially different from the drawing of normal children. It is not that she had an accelerated development in this sphere but rather that her development was totally anomalous. Even her earlier

drawings showed few of the properties associated with infant drawings . . . Perspective, for instance, was present from the start.[8]

These drawings of Nadia's, I now suggest, bear astonishing parallels to high cave art.

Figure 8 shows part of the big horse panel from Chauvet, Figure 9 a drawing of horses made by Nadia—one of her earliest—at age three years five months. Figure 10 shows a tracing of horses from Lascaux, Figure 11 another of Nadia's early drawings. Figure 12 shows an approaching bison from Chauvet, Figure 13 an approaching cow by Nadia at age four. Figure 14 shows a mammoth from Pech Merle, Figure 15 two elephants by Nadia at age four. Figure 16 a detail of a horsehead profile from Lascaux, Figure 17 a horsehead by Nadia at age six. Figure 18, finally, a favourite and repeated theme of Nadia's, a rider on horseback, this one at age five.

The remarkable similarities between the cave paintings and Nadia's speak for themselves. There is first of all the striking naturalism and realism of the individual animals. In both cases, as Jean Clottes writes of the Chauvet paintings, 'These are not stereotyped images which were transcribed to convey the concept "lion" or "rhinoceros", but living animals faithfully reproduced.'[9] And in both cases, the graphic techniques by which this naturalism is achieved are very similar. Linear contour is used to model the body of the animals. Foreshortening and hidden-line occlusion are used to give perspective and depth. Animals are typically 'snapped' as it were in active motion—prancing, say, or bellowing. Liveliness is enhanced by doubling up on some of the body contours. There is a preference for side-on views. Salient parts, such as faces and feet, are emphasized—with the rest of the body sometimes being ignored.

Yet it is not only in these 'sophisticated' respects that the cave drawings and Nadia's are similar, but in some of their more idiosyncratic respects, too. Particularly notable in both sets of drawings is the tendency for one figure to be drawn, almost haphazardly, on top of another. True, this overlay may

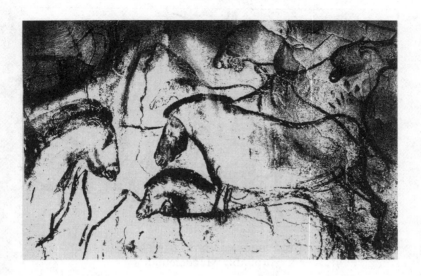

Fig. 8. Painted horses from Chauvet Cave (Ardèche),
probably Aurignacian

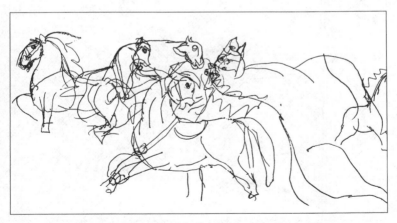

Fig. 9. Horses by Nadia, at 3 years 5 months

Fig. 10. Painted and engraved horses from Lascaux (Dordogne), probably Magdalenian

Fig. 11. Horses by Nadia, at 3 years 5 months

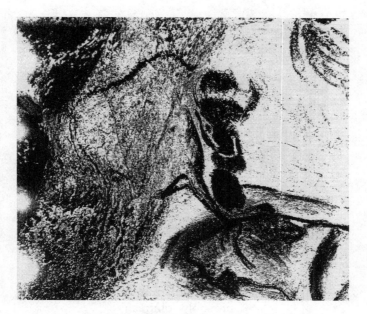

Fig. 12. Painted bison from Chauvet cave (Ardèche),
probably Aurignacian

Fig. 13. Cow by Nadia, at approximately 4 years

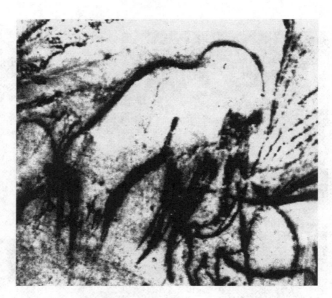

Fig. 14. Painted mammoth from Pech Merle (Lot),
probably Solutrean

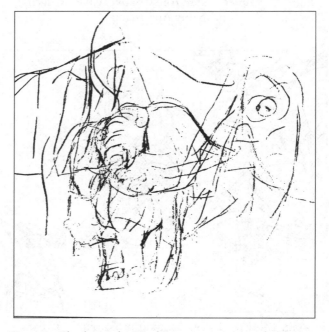

Fig. 15. Elephants by Nadia, at approximately 4 years

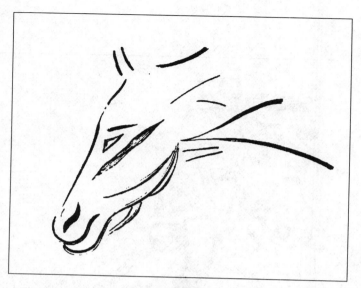

Fig. 16. Engraved horsehead from Lascaux (Dordogne),
probably Magdalenian

Fig. 17. Horsehead by Nadia, at approximately 6 years

Fig. 18. Horse and rider by Nadia, at 5 years

sometimes be interpretable as a deliberate stylistic feature. Clottes, for example, writes about Chauvet: 'In many cases, the heads and bodies overlap, doubtless to give an effect of numbers, unless it is a depiction of movement.'[10] In many other cases, however, the overlap in the cave paintings serves no such stylistic purpose and seems instead to be completely arbitrary, as if the artist has simply paid no notice to what was already on the wall. And the same goes for most of the examples of overlap in Nadia's drawings. Figure 19, for example, shows a typical composite picture made by Nadia at age five—comprising a cock, a cat, and two horses (one upside down).

In Nadia's case, this apparent obliviousness to overlap—with the messy superimpositions that resulted—may in fact have been a positive feature of her autism. Autistic children have often been noted to be unusually attentive to detail in a sensory array, while being relatively uninfluenced—and even maybe unaware of—the larger context.[11] Indeed, such is their tendency to focus on parts rather than wholes that, if and when the surrounding context of a figure is potentially misleading or confusing, they may actually find it easier than normal people to ignore the context and see through it. Amitta Shah and Uta Frith have shown, for example, that autistics perform quite exceptionally well on the so-called 'hidden figure' test, where the task is to find a target figure that has been deliberately camouflaged by surrounding lines.[12]

There is no knowing whether the cave artists did in fact share with Nadia this trait which Frith calls 'weak central coherence'.[13] But if they did do so, it might account for another eccentricity that occurs in both series of drawings. Selfe reports that Nadia would sometimes use a detail that was already part of one figure as the starting point for a new drawing—which would then take off in another direction—as if she had lost track of the original context.[14] And it seems (although I admit this is my own post hoc interpretation) that this could even happen halfway through, so that a drawing that began as one kind of animal would turn into another. Thus Figure 20 shows a strange composite animal produced by Nadia, with

Fig. 19. Superimposed animals by Nadia, at 6 years 3 months

Fig. 20. Composite animal, part giraffe, part donkey,
by Nadia, at approximately 6 years

the body of giraffe and the head of donkey. The point to note is that chimeras of this kind are also to be found in cave art. The Chauvet cave, for example, has a figure that apparently has the head of a bison and the trunk and legs of a man.

What lessons, if any, can be drawn from these surprising parallels? The right answer might, of course, be: None. I am sure there will be readers—including some of those who have thought longest and hardest about the achievements of the Ice Age artists—who will insist that all the apparent resemblances between the cave drawings and Nadia's can only be accidental, and that it would be wrong—even impertinent—to look for any deeper meaning in this 'evidence'. I respect this possibility, and agree we should not be too quick to see a significant pattern where there is none. In particular, I would be the first to say that resemblances do not imply identity. I would not dream of suggesting, for example, that the cave artists were themselves clinically autistic, or that Nadia was some kind of a throwback to the Ice Age. Yet, short of this, I still want to ask what can reasonably be made of the parallels that incontrovertibly exist.

To start with, I think it undeniable that these parallels tell us something important about what we should *not* assume about the mental capacities of the cave artists. Given that Nadia could draw as she did *despite* her undeveloped language, impoverished cognitive skills, apparent lack of interest in communication, and absence of artistic training, it is evident that so too *could* the cave artists have done so. Hence the existence of the cave drawings should presumably *not* be taken to be the proof, which so many people have thought it is, that the cave artists had essentially modern minds. Tattersall, for instance, may claim that '[Chauvet] dramatically bolsters the conclusion that the first modern people arrived in Europe equipped with all of the cognitive skills that we possess today';[15] but he is clearly on less solid ground than he supposes.

Next—and I realize this is bound to be more controversial—I think it possible that the parallels also tell us something more positive about what we *can* assume about the artists' minds. For suppose it were the case that Nadia could draw as she did *only because* of her undeveloped language and other impoverishments. Suppose, indeed, it were more generally the case that a person not only *does not need* a typical modern mind to draw like that but *must not have* a typical modern mind to draw like that. Then the cave paintings might actually be taken to be proof positive that the cave artists' minds were essentially pre-modern.

In Nadia's case, there has in fact already been a degree of rich speculation on this score: speculation, that is, as to whether her drawing ability was indeed something that was 'released' in her only because her mind failed to develop in directions that in normal children more typically smother such ability. Selfe's hypothesis has always been that it was Nadia's language—or rather her failure to develop it—that was the key.

At the age of six years Nadia's vocabulary consisted of only ten one-word utterances, which she used rarely. And, although it was difficult to do formal tests with her, there were strong hints that this lack of language went along with a severe degree of literal-mindedness, so that she saw things merely as they appeared at the moment and seldom if ever assigned them to higher-level categories. Thus,

it was discovered that although Nadia could match difficult items with the same perceptual quality, she failed to match items in the same conceptual class. For example, she could match a picture of an object to a picture of its silhouette, but she failed to match pictures of an armchair and a deck chair from an array of objects that could be classified on their conceptual basis.[16]

It was this very lack of conceptualization, Selfe believes, that permitted Nadia to register exactly how things looked to her. Whereas a normal child of her age, on seeing a horse, for example, would see it—and hence lay down a memory of it—as a token of the category 'horse', Nadia was simply left with the original visual impression it created.

Selfe went on to examine several other autistic subjects who also possessed outstanding graphic skills (although none, it must be said, the equal of Nadia), and she concludes that for this group as a whole the evidence points the same way.

It is therefore proposed that without the hypothesised domination of language and verbal mediation in the early years when graphic competence was being acquired, these subjects were able to attend to the spatial characteristics of their optic array and to represent these aspects in their drawing . . . These children therefore have a more direct access to visual imagery in the sense that their drawings are not so strongly 'contaminated' by the usual 'designating and naming' properties of normal children's drawings.[17]

Thus, whereas a normal child when asked to draw a horse would, in the telling words of a five-year-old, 'have a think, and then draw my think', Nadia would perhaps simply have had a look at her remembered image and then drawn that look.

This hypothesis is, admittedly, somewhat vague and open-ended; and Selfe herself considers it no more than a fair guess as to what was going on with Nadia. However, most subsequent commentators have taken it to be at least on the right lines, and certainly nothing has been proposed to better it. I suggest, therefore, we should assume, for the sake of argument at least, that it is basically correct. In which case, the question about the cave artists immediately follows. Could it be that in their case, too, their artistic prowess was due to the fact that they had little if any language, so that their drawings likewise were uncontaminated by 'designating and naming'?

There are two possibilities we might consider. One is that language was absent in the general population of human beings living in Europe 30,000 years ago. The other is that there were at least a few members of the population who lacked language and it was from amongst this subgroup that all the artists came. But this second idea—even though there is no reason to rule it out entirely (and though the philosopher Daniel Dennett tells me it is the one he favours)—would seem to involve too much special pleading to deserve taking further, and I suggest we should focus solely on the first.

Then, we have to ask: is it really in any way plausible to suppose that human beings of such a relatively recent epoch had as yet not developed the capacity for full-scale language? The standard answer, coming from anthropology and archaeology, would certainly be: No. Human spoken language surely had its beginnings at least a million years ago, and most likely had already evolved to more or less its present level by the time the ancestral group of Homo sapiens sapiens left Africa around 100,000 years ago. By the date of the first cave paintings, therefore, there can be no question of there being any general deficiency in people's capacity to name or designate.

Yet there are revisionist ideas about this in the air. Everybody agrees that *some* kind of language for *some* purpose has likely been in existence among humans for most of their history since they parted from the apes. But Robin Dunbar, for example, has argued that human language evolved originally not as a general purpose communication system for talking about anything whatever, but rather as a specifically social tool for negotiating about—and helping maintain—interpersonal relationships.[18] And Steven J. Mithen has taken up this idea and run with it, arguing that the 'linguistic module' of the brain was initially available only to the module of 'social intelligence', not to the modules of 'technical intelligence' or 'natural history intelligence'.[19] So, to begin with, people would—and could—use language only as a medium for naming and talking about other people and their personal concerns, and not for anything else.

Even so, this idea of language having started off as a sub-speciality may not really be much help to the argument at hand. For Mithen himself has argued that the walls around the mental modules came down at the latest some 50,000 years ago. In fact, he himself takes the existence of the supposedly 'symbolic' Chauvet paintings to be good evidence that this had already happened by the date of their creation: 'All that was needed was a connection between these cognitive processes which had evolved for other tasks to create the wonderful paintings in Chauvet Cave.'[20] Therefore, other things being equal, even Mithen could not be expected to countenance the much later date that this line of thinking that stems from Nadia indicates.

However, suppose that while Mithen is absolutely right in his view of the sequence of changes in the structure of the human mind, he is still not sufficiently radical in his timing of it. Suppose that the integration of modules that he postulates did not take place until, say, just 20,000 years ago, and that up to that time language did remain more or less exclusively social. So that the people of that time—like Nadia today—really did not have names for horses, bison, and lions (not to mention chairs). Suppose, indeed, that the very idea of something representing 'the bison as such' had not yet entered their still evolving minds. Then, I suggest, the whole story falls in place.

J. M. Keynes wrote of Isaac Newton that his private journals and notebooks reveal him to have been not the first scientist of the Age of Reason but the last of the magicians.[21] Now, likewise, we might say that the cave paintings reveal their makers to have been not the first artists of the age of symbolism but the last of the innocents.

But 20,000 years ago? No language except for talking about other people? In an experiment I did many years ago, I found clear evidence that rhesus monkeys are cognitively biased towards taking an interest in and making categorical distinctions between *other rhesus monkeys*, while they ignore the differences between individuals of *other species*—cows, dogs, pigs, and so on.[22] I am therefore probably more ready than most to believe that early humans might have had minds that permitted them to think about other people in ways quite different from the ways they were capable of thinking about non-human animals. Even so, I too would have thought the idea that there could still have been structural constraints on the scope of human language until just 20,000 years ago too fantastic to take seriously, were it not for one further observation that seems to provide unanticipated confirmation of it. This is the striking difference in the representation of humans as opposed to animals in cave art.

Note that the hypothesis, as formulated, makes a testable prediction. If before 20,000 years ago people had names available for talking about other human individuals but not for other animals, and if it were indeed this lack of naming that permitted those artists to depict animals so naturalistically, then this naturalism ought *not* to extend to other human beings. In other words, representations of humans should either be missing altogether from the cave paintings, or if present should be much more stereotypical and modern.

But, behold, this is exactly what *is* the case. As a matter of fact, there are no representations of humans at Chauvet. And when they do occur in later paintings, as at Lascaux at 17,000 years ago, they are nothing other than crudely drawn iconic symbols. So that we are presented in a famous scene from Lascaux, for example, with the conjunction of a well-modelled picture of a bison with a little human stick-figure beside it (Fig. 21). In only one cave, La Marche, dating to 12,000 years ago, are there semi-realistic portrayals of other humans, scratched on portable plaquettes—but even these appear to be more like caricatures.

Nadia provides a revealing comparison here. Unlike the cave artists, Nadia as a young girl had names neither for animals nor people. It is to be expected, therefore, that Nadia, unlike the cave artists, would in her early drawings have accorded both classes of subject equal treatment. And so she did. While it is true that Nadia drew animals much more frequently than people, when she did try her hand at the latter she showed quite similar skills. Nadia's pictures of footballers and horsemen at age five, for example, were as natural-looking as her pictures of horses themselves. Figure 22 shows Nadia's drawing of a human figure, made at age four.

I accept, of course, that none of these comparisons add up to a solid deductive argument. Nonetheless, I think the case for supposing that the cave artists did share some of Nadia's

Fig. 21. Painted bison and human figure, Lascaux
(Dordogne), probably Magdalenian

Fig. 22. Human figure by Nadia, at approximately 4 years

mental limitations looks surprisingly strong. And strong enough, surely, to warrant the question of how we might expect the story to continue. What would we expect to have happened—and what did—when the descendants of those early artists finally acquired truly modern minds? Would we not predict an end to naturalistic drawing across the board?

In Nadia's case it is significant that when at the age of eight and more, as a result of intensive teaching, she did acquire a modicum of language, her drawing skills partly (though by no means wholly) fell away. Elizabeth Newson, who worked with her at age seven onwards, wrote:

Nadia seldom draws spontaneously now, although from time to time one of her horses appears on a steamed up window. If asked, however, she will draw: particularly portraits . . . In style [these] are much more economical than her earlier drawings, with much less detail . . . The fact that Nadia at eight and nine can produce recognisable drawings of the people around her still makes her talent a remarkable one for her age: but one would no longer say that it is *unbelievable.*[23]

So, Newson went on, 'If the partial loss of her gift is the price that must be paid for language—even just enough language to bring her into some kind of community of discourse with her small protected world—we must, I think, be prepared to pay that price on Nadia's behalf.'

Was this the story of cave art, too? With all the obvious caveats, I would suggest it might have been. What we know is that cave art, after Chauvet, continued to flourish with remarkably little stylistic progression for the next twenty millennia (though, interestingly, not without a change occurring about 20,000 years ago in the kinds of animals represented).[24] But then at the end of the Ice Age, about 11,000 years ago, for whatever reason, the art stopped. And the new traditions of painting that emerged over five millennia later in Assyria and Egypt were quite different in style, being much more conventionally childish, stereotyped and stiff. Indeed, nothing to equal the naturalism of cave art was seen again in Europe until the Italian Renaissance, when lifelike perspective drawing was reinvented, but now as literally an 'art' that had to be learned through long professional apprenticeship.

Maybe, in the end, the loss of naturalistic painting was the price that had to be paid for the coming of poetry. Human beings could have Chauvet or the *Epic of Gilgamesh* but they could not have both. I am sure such a conclusion will strike many people not merely as unexpected but as outlandish. But then human beings are a great miracle, and if their history were not in some ways unexpected and outlandish they would be less so.

POSTSCRIPT

When this essay was published in the *Cambridge Archaeological Journal*, it was followed by eight critical commentaries by archaeologists and psychologists.[25] In my reply to these commentaries, I was able to expand on and clarify some of my original points. This is an abbreviated version of that reply (with the thrust of the commentators' criticisms emerging in the course of my responses to them).

Paul Bahn remarks that one of the joys of being a specialist in prehistoric art is the stream of strange ideas that come his way. I should say that one of the joys of being a non-specialist is to have an opportunity such as this to be listened to, enlarged upon, and corrected by scholars who know the field better than I do.

The main purpose of my essay was to challenge the over-confident 'modernist' (with a small 'm') interpretation of Ice Age art (not to mention other aspects of Upper Palaeolithic culture) that pervades both academic and popular accounts. As Robert Darnton has written in a different context, 'Nothing is easier than to slip into the comfortable assumption that Europeans [of the past] thought and felt just as we do today—allowing for the wigs and wooden shoes'.[26] And equally I would say nothing is easier than to slip into the assumption that the Ice Age artists created pictures in the way and even for some of the reasons that modern artists do today—allowing for the reindeer picks and tallow candles. 'We constantly need to be shaken out of a false sense of familiarity with the past,' Darnton continued, 'to be administered doses of culture shock.'

What I set out to demonstrate in the first part of the essay was the shocking truth that there are quite other ways of being an artist than the one we take for granted. Nadia's skill was such that, if we did not know the provenance of her drawings, we might well assume that they came from the hand of someone with all the promise of a young Picasso. Yet Nadia was mentally disabled. She lacked the capacity to speak or to symbolize, and she created her art only for her own amusement. I argued, therefore, that just as we might so easily misinterpret Nadia's drawings if we were to come across them cold, so there is the possibility that we may have already been misinterpreting cave art. At the very least, scholars should be more cautious than they have been before jumping to grandiose conclusions about the mentality of the Ice Age artists.

Now, to this, the negative argument of the paper about what we should *not* conclude about cave art, two kinds of objection are raised.

The first consists in denying that there is in fact any significant similarity between cave art and Nadia's. Paul Bahn claims he simply cannot see the similarity. Ezra Zubrow thinks it might be due to selective sampling, or else merely chance. Steven Mithen has reservations about the drawing techniques and says he sees 'a glaring difference in the quality of line: Nadia appears to draw in a series of unconnected lines, often repeated in the manner of a sketch, while the dominant character of cave art is a confidence in line, single authoritative strokes or engraved marks'.

But, as Selfe's description of Nadia's technique makes clear, Mithen is making more of this technical difference than is warranted. 'Nadia used fine, quickly executed lines. Her motor control was highly developed . . . Her lines were firm and executed without unintentional wavering . . . She almost invariably appeared to have a definite idea about what she was drawing so there were no wasted lines.'[27] And if Bahn and Zubrow think the overall similarities are nonexistent or accidental, all I can say is: look again. Or, better, do not rely on the few illustrations of this paper, but take the hundred or so drawings by Nadia in Selfe's book and match them however you will with works from Chauvet or Lascaux. Mithen comments on 'the remarkable continuity in subject matter and style of Upper Palaeolithic art'. I defy anyone with an eye for style not to see how easily Nadia's drawings could have been part of this same tradition (indeed, how they are in some ways closer to Chauvet on one side and Lascaux on the other, than Chauvet is to Lascaux!).

It is not fair, perhaps, to play the connoisseur and question the aesthetic sensitivity of those who will not see things my way. But I confess that, when Bahn asks why I make so much of Nadia in my paper as against other savant artists such as Stephen Wiltshire,[28] and implies that Stephen Wiltshire's drawings would have made an equally good (or, as he thinks, bad) comparison for cave art, it does make me wonder about the quality of his critical judgment. For I'd say it should be obvious to anyone with a good eye that Nadia's drawings of animals *demand* this comparison, whereas Stephen Wiltshire's drawings of buildings simply do not.

The second kind of objection to the negative argument about what we should not conclude about cave art is at a different level. It consists in claiming that what this argument does is to treat cave art as if it were an isolated phenomenon, whereas it ought properly to be considered in the context of the rest of the surrounding culture. Steven Mithen and Ian Tattersall, for example, both concede that the argument about cave art being produced by minds similar to Nadia's might possibly go through, *if* the art was all we had to go on. But, they say, when the achievements of Upper Palaeolithic culture are considered as a whole, this interpretation simply does not wash. There is too much else in the archaeological record that speaks to the presence at this time of sophisticated, symbol-using, language-saturated minds: evidence of body decoration, music, funerary rituals, elaborate trade networks, and so on. If human beings were so far advanced in all these other areas, surely they must have been using the same high-level mental skills in their art also.

This sounds persuasive, until we realize that it largely begs the question. I'd agree it might be unarguable that, if it were certainly established that these other cultural activities really occurred in the way that archaeologists imagine and involved those high-level skills, then it would follow that art did too. But what makes us so sure that Upper Palaeolithic humans *were* engaging in ritual, music, trading, and so on *at the level that everyone assumes*? One answer that clearly will not do here is to say that these were the same humans who were producing symbolic art! Yet, as matter of fact this is just the answer that comes across in much of the literature: cave art is taken as the first and best evidence of there having been a leap in human mentality at about this time, and the rest of the culture is taken as corroborating it.

Of course, another answer might be that high-level symbolic thought had to be involved in these other activities because when we ourselves engage in similar activities today, we use our full range of mental skills to do so. But, again, this is precisely the kind of logic I (and Robert Darnton above) mean to question. Just as there are ways of drawing beautiful and complex pictures that are not our ways, we should be alert to the possibility that there are ways of having intense and meaningful social engagements that are not our ways—including forms, though not exactly our forms, of trade, ritual, dance, and so on. In particular, we should not assume any necessary role in any of these things for universal, cross-domain language.

I turn now to the positive argument that I mounted in the second half of the essay, about what perhaps we *can* conclude about cave art: namely, that the people who produced it not only might not have had modern minds like ours, but really *did not*—and in particular, that they did still have minds more like Nadia's, with underdeveloped language. I am hardly surprised that this suggestion has met with more scepticism and hostility than the first, and indeed that Daniel Dennett is virtually alone among the commentators in looking kindly on it—for it is of course closer to the kind of no-holds-barred 'what if?' speculation that philosophers are familiar with than it is to normal science.

But there may be another reason why Dennett likes this argument, while others do not. For I realize now that there has been a general misunderstanding of my position, one that I did not see coming, but which if I had seen I should have tried to head off earlier. It appears that almost all the other commentators have taken it for granted that when I talk about the difference between a pre-modern and modern mind (or a linguistically restricted/unrestricted mind, or a cognitively rigid/fluid mind), I must be talking about a genetically determined difference in the underlying brain circuitry. That's to say, that I must be assuming that humans were in the past *born* with a pre-modern mind, while today they are (except for unfortunate individuals such as Nadia) *born* with a modern mind.

But this not my position at all. For, in line with Dennett's own ideas about recent cognitive evolution,[29] I actually think it much more likely that the change from pre-modern to modern came about not through genetic changes in innately given 'hardware' but rather through environmental changes in the available 'software': in other words, I think that pre- modern humans became modern humans when their environment— and specifically the linguistic and symbolic environment inherited through their culture—became such as to reliably programme their minds in quite new ways.

In the longer run, of course, there must also have been important genetic changes. No modern environment could make a modern human of a chimpanzee or even of one of our ancestors from, say, 100,000 years ago. Still, I'd suggest that, over the time period that concerns us here, genetic changes in the structure of the brain actually account for very little. It is primarily the modern twentieth-century cultural environment that makes modern humans of our babies today, and it was primarily the pre-modern Upper Palaeolithic environment that made pre-modern humans of their babies then (so that, if our respective sets of babies were to swap places, so would their minds).

Now, I realize that in this regard the analogy I drew with Nadia and with autism is potentially misleading. For, as Bahn does well to point out, Nadia like most autistic children almost certainly had some kind of congenital brain abnormality (although the evidence is unclear as to whether, as Bahn claims, there was specific damage to her temporal lobes). Unlike the pre-modern humans we are talking about, Nadia did not have underdeveloped software, but rather she had damaged hardware. I ought to have made it clear, therefore, that the similarity I see between Nadia and pre-modern humans is at the level of the functional architecture of their minds rather than of the anatomy of their brains. Specifically, both pre- modern humans (because of their culture) and Nadia (because of her brain damage) had very limited language, and in consequence both had heightened pictorial memory and drawing skills.

I hope it will be obvious how, with this being the proper reading of my argument, some of the objections of the commentators no longer strike home. In particular there need be no great problem in squaring my suggestion about the relatively late arrival of modern minds in Europe with the known facts about the geographic dispersion of the human population. Paul Bloom and Uta Frith rightly observe that a genetic trait for modernity cannot have originated in Europe as late as I suggest and subsequently spread through the human population, because in that case there is no way this trait could have come to be present in the Australian aborigines whose ancestors moved to Australia 50,000 years ago. Steven Mithen is worried by the same issue and reckons the only answer (by which he is clearly not convinced) is that there might have been convergent evolution. But if the change from pre-modern to modern resulted from a change in the cultural environment rather than in genes, then, wherever this cultural development originated, it could easily have spread like wildfire in the period between say 20,000 and 10,000 years ago—right the way from Europe to Australia, or, equally possibly, from Australia to Europe.

There are, however, other important issues that I still need to address. First, there is the question of whether there really is any principled connection between graphic skills and lack of language. Several commentators note that lack of language is certainly not sufficient in itself to 'release' artistic talent, and indeed that the majority of autistic children who lack language do not have any such special talent at all. But this is hardly surprising and hardly the issue. The issue is whether lack of language is a necessary condition for such extraordinary talent to break through. And here the evidence is remarkably and even disturbingly clear: '*no normal preschool child has been known to draw naturalistically.* Autism is apparently a necessary condition for a preschool child to draw an accurate detail of natural scenes.'[30]

While it is true that all known artistic savants have in fact been autistic, I agree with Chris McManus, and indeed it is an important part of my argument, that autism as such is probably not the relevant condition. Rather, what matters primarily is the lack of normal language development that is part and parcel of the syndrome. I stressed in my essay the fact that when Nadia did at last begin to acquire a little language at eight years old, her graphic skills dramatically declined. If Nadia were alone in showing this pattern, it might not mean much. But in fact it seems to be the typical pattern—in so far as anything is typical—of other children who have shown similar artistic talents at a very young age. And it provides strong corroborative evidence for the idea that language and graphic skills are partly incompatible.

Bahn is right to point out that there have been exceptions to this general rule. But he is far from right to hold up the case of Stephen Wiltshire as a knock-down counter-example. As I mentioned above, Stephen Wiltshire's drawings are so different in style from cave art that I would never have thought to discuss them in the present context. But, as Bahn makes so much of Stephen Wiltshire's case, I should relay a few of the relevant facts.[31]

Stephen Wiltshire, like Nadia, was severely autistic as a child and failed to develop language normally. He began to produce his drawings at the age of seven, whereas Nadia began earlier at age three. But like Nadia, Stephen still had no language when this talent first appeared. At age nine, however, he did begin to speak and understand a little. And it is true that, in contrast to Nadia, Stephen's artistic ability thereafter grew alongside his language rather than declined. But what makes his case so different from Nadia's is that Stephen, who was much less socially withdrawn than Nadia, was *intensively coached by an art teacher from the age of eight onwards*. There is every reason to think, therefore, that the continuation of his ability into adolescence and adulthood was not so much the persistence of savant skills, as the replacement of these skills by those of a trained artist.

Although Bahn quotes me as suggesting that 'a person must not have a typical modern mind to draw like that', he must realize it is no part of my argument to claim that no person with full possession of language can *ever* draw naturalistically—*even with training*. How could I possibly claim this—given the obvious presence in the contemporary world of countless people with language who have indeed *learned* to draw perfectly well? Rather, the 'draw like that' in my statement clearly refers to the ability to draw *like Nadia*—in other words, spontaneously, without formal training or access to the canon of tricks we learn in art school. The point is that for normal people this ability *never* comes that easily. As Gombrich has written, 'this imitation of visual reality must be very complex and indeed a very elusive affair, for why should it otherwise have taken so many generations of gifted painters to learn its tricks'.[32] But in Nadia's case, the imitation of visual reality seems, by contrast, to have been very simple and direct.

Returning to the issue of why Nadia's skills declined, Bahn speculates that a more plausible explanation than the advent of language is the death of Nadia's mother at about the same time. But Bahn fails to acknowledge that neither of the psychologists who actually worked with Nadia and her family considered this a likely explanation. Nor does he mention (presumably because it wouldn't suit) that Stephen Wiltshire also had a parent die, his father: but in his case the death occurred at the beginning of his drawing career rather than the end of it.

Given that savant skills generally do come to an end, unless perhaps as in Stephen Wiltshire's case there is active intervention by a teacher, is there really any parallel for this in the history of art? On this question I regret that, in the flourish of the final paragraphs, I oversimplified a story that in reality has several complex strands. It is true, as I stated, that at end of the last Ice Age, the tradition of cave art in the Franco-Cantabrian region where it had flourished for the previous twenty millennia came to a surprising end. But Bahn is right to take me to task for not acknowledging the persistence of rock paintings elsewhere, and especially the newer tradition that took off in southern Spain about 11,000 years ago and which seems to have links with African art down to nearly the present day.

These later paintings from the Spanish Levant and Africa are so different both in content and style from the ones we have been discussing that I have no hesitation in reasserting that 'the art stopped'. But I am still somewhat embarrassed that Dennett should take this to be 'the critical piece of evidence' in favour of my theory. For I agree I was exaggerating when I wrote that naturalistic painting died out altogether in Europe at the end of the Ice Age, until it was reinvented in the recent Middle Ages. There are certainly fine examples of naturalism to be found in Spanish-Levantine rock art, and, from a later period, in Greek vase painting and Roman murals (and, further afield, in the rock art of the San bushmen.)

Yet, what kind of examples are these, and what do they tell us? I think it undeniable that, for all their truth to visual reality, they are still relatively formulaic and predictable: copybook art that lacks the extraordinary freshness of vision that makes us catch our breath on first seeing Chauvet or Lascaux—as Newson said of Nadia's post-language drawings, 'remarkable but no longer unbelievable'. And if they have that copybook feel to them, I expect that is because that is really what they are: already we are into the modern era where learned tricks of artistry are having to substitute for the loss of the innocent eye.

I avoided any discussion in my essay of the motivations—individual or social—for creating Ice Age art. But none of the commentators on the paper have been so cautious. And since Bloom and Frith's observations, especially, are provocative, let me join in finally with my own pennyworth.

Nadia, it seems, drew for the sake of her own pleasure in the drawing. 'She drew intensively for varying intervals of time but not for more than one minute . . . After surveying intently what she had drawn she often smiled, babbled and shook her hands and knees in glee.'[33] But she had no interest in sharing her creation with anyone else. And, as Bloom and Frith point out, it is characteristic of autistic artists generally that 'they produce, but do not *show*'.

This prompts these authors to continue: 'It is interesting to speculate about a species, different from modern humans, that did not have ostensive communication, yet was able to outperform them in artistic production.' But, though I doubt this is what they had in mind, the fact is we already know of many other species that come close to doing just what they suggest: in other words, they produce 'artistic displays' without any insight into what they are doing or why they do it and without any conscious intention to communicate. And the place where it happens most dramatically and obviously is in the context of *courtship and sexual advertisement*. The nightingale with its song, the peacock with its tail, the octopus with its dance . . . True, in such cases the aesthetically brilliant display is at some level meant to impress another individual; but the communication is certainly not ostensive or consciously thought out—rather, it is species-typical behaviour that has evolved by *sexual selection* as a way by which the artist is able to signal his or her quality to a prospective mate.

Sexual selection is increasingly being recognized by human biologists as having been a potent factor in human evolution. Geoffrey Miller believes that there is hardly any aspect of human skilled performance that has not been profoundly influenced by the exigencies of mate choice.[34] And Steven Mithen has recently speculated that the main use of Acheulian hand axes by early humans may have been by males to woo females, with the axe being a reliable token of the axe-maker's skills.[35] I would suggest it is quite possible that cave art evolved in this context as well: with painting after painting being produced by fired-up young men (probably men, but possibly women too) as an implicit demonstration of the artists' potential qualities as sires and parents.

Would this be 'art for art's sake', as some of the first theorists of cave art argued? Not quite. But it would be art, stemming from the soul and body of the artist, offered like the song of a bird in celebration of a mystery, without the artist needing to be in any way aware of how *his own sake* was being served.

13

Scientific Shakespeare

The year 1987 being the three hundredth anniversary of Newton's *Principia Mathematica*, published in 1687, I have been considering some of the other anniversaries that fall in the same year: Chaucer's *Canterbury Tales* (1387), Leonardo's painting of the *Virgin of the Rocks* (1487), Marlowe's *Tamburlaine* (1587), Mozart's *Don Giovanni* (1787), Nietzsche's *Genealogy of Morals* (1887).

The eighty-seventh year of the century would seem to have been an auspicious time for art and scholarship. Yet, if we had to choose just one of these great works to celebrate, we ought surely to give first place to the *Principia*. Newton's law of gravitation, which states that 'every body in the universe attracts every other with a force inversely proportional to the square of the distance between them', has been described as the greatest generalization ever made by the human mind.

What, then, if we had to choose just one of these works *to consign to oblivion*? If the choice were forced, we ought surely to *let go* the *Principia*. How so? Because, of all those works, Newton's was the only one that would have been replaceable. Quite simply: if Newton had not written his book, then someone else would have written another book just like it—probably within the space of a few years.

C. P. Snow, in *The Two Cultures*, extolled the great discoveries of science as 'scientific Shakespeare'.[1] But in one way he was fundamentally mistaken. Shakespeare's plays were Shakespeare's plays and no one else's; scientific discoveries, by contrast, belong—ultimately—to no one in particular. Take away the work of the *person* Shakespeare, or Chaucer, or Mozart and you would take away the contingent creation of a one-off human mind; take away the work of Newton, or Darwin, or Einstein and you would take away nothing that could not eventually be replaced by Mind at large.

It may be unfashionable to say that the job of science is to uncover God's pre-existing truths. But, notwithstanding today's 'subject-centred theories of reality', I think science does just that. There *are* pre-existing truths out there waiting to be found, and it *is* the job of the scientist to uncover them. In no way, however, can the same be said of art. There are no pre-existing novels out there waiting to be written, nor pre-existing pictures waiting to be painted.

Consider the disputes that arise in science, but not in art, about 'priority'. Newton quarrelled fiercely with Leibniz about which of them had in fact invented the differential calculus before the other, and with Hooke about who had discovered the inverse square law. But while, say, there may once have been room for dispute about whether Marlowe actually wrote Shakespeare's plays, no one would ever have suggested that Marlowe got there *before* Shakespeare.

Newton had a dog called Diamond. One day, the story goes, the dog knocked over a candle, set fire to some papers and destroyed 'the unfinished labours of some years'. 'Oh Diamond, Diamond!' Newton cried, 'thou little knowest the mischief done!'[2] Suppose that the papers had been the manuscript of the *Principia*, and that Newton, in chagrin or despair, had given up doing science. Mischief, indeed. Nonetheless, Diamond's mischief would hardly have changed the course of history. Imagine however that Diamond had instead been Chaucer's dog and that he had set fire to the *Canterbury Tales*. The loss would truly have been irrecoverable.

General James Wolfe said of Gray's *Elegy*, 'I would rather have written that poem than take Quebec'. In 1887 the architect Gustave Eiffel built the Eiffel Tower. Would it be understandable for anyone to say he would rather have built the Eiffel Tower than have written the *Principia*? It would depend on what his personal ambitions were. The *Principia* was a glorious monument to human intellect, the Eiffel Tower was a relatively minor feat of romantic engineering. Yet the fact is that while Eiffel did it *his* way, Newton merely did it God's way.

To be God's scribe, as Newton was, is not to have an undistinguished role. Nevertheless I would conclude—until someone shows me why I am wrong—that any person who wants to leave their own peculiar mark on the landscape of other people's minds should build towers, paint pictures, or write stories rather than devote themselves to uncovering the scientific truth. As a scientist myself I find this conclusion worth thinking about, if not worth trying to live by.

14

The Deformed Transformed

And Jesus said, How hardly shall they that have riches enter into the kingdom of God! For it is easier for a camel to go through a needle's eye, than for a rich man to enter into the kingdom of God. And they that heard it said, Who then can be saved? And he said, The things which are impossible with men are possible with God . . . There is no man that hath left house, or parents, or brethren, or wife, or children, for the kingdom of God's sake, Who shall not receive manifold more in this present time, and in the world to come life everlasting.

Luke 18: 24–30

In 1711, William Derham, Canon of Windsor and Fellow of the Royal Society, gave a series of lectures surveying the physical and natural world, with the object of demonstrating how perfect is God's Creation in every respect. His survey, published as his *Physico-Theology*, ranged from the existence of gravity (without which, Derham observed, the earth would fly apart) to the distribution of venomous snakes (of which, he noted with evident satisfaction, all the worst ones live in lands peopled by non-Christians). In particular, Derham found that man himself is, 'every Part of him, every Thing relating to him contriv'd, and made in the very best Manner; his Body fitted up with the utmost Foresight, Art and Care'.[1] Indeed, if anyone were to suggest, to the contrary, that there are ways in which the design of human beings can be *improved*, it would be blasphemy.

That was three hundred years ago. Times have moved on, and our style of argument has changed. Nonetheless, as modern-day followers of Darwin we remain no less committed than

Derham and his fellow philosopher-theologians to an idea of optimal design in nature. We may no longer believe that we live in the best of all *possible* worlds. But we do have reason to believe that we live in the best—or close to best—of all *available* worlds.

It is easy to see why. Let's suppose we are considering the evolution of some desirable phenotypic trait that can be scaled from *less* to *more*—intelligence, say, or beauty. Then theory tells us that, if and when an increase in this trait is indeed an available option within the biological 'design space' of the species, and if it will indeed bring an increase in fitness to go that way, then this is the way that natural selection will drive the population as a whole. In other words, it is highly likely that, provided there is time and opportunity, the species will evolve to the point where this trait goes to fixation at the best level throughout the population.

We may well assume, then, just as Derham did, that in general there will be little if any room for making further progress, at least by natural means. Nature will already have done for human beings the very best that in practice can be done. The last thing we shall expect, therefore, is that any significant improvement in bodily or mental capacities can be achieved as a result simply of minor tinkering with the human design plan.

Yet the truth is that there is accumulating evidence to suggest just this.

For a start, as medical science grows ever bolder, it is proving to be a relatively straightforward task for doctors to increase human performance levels by direct intervention in human embryology and later physiology—with, say, foetal androgens for brain growth, anabolic steroids for strength, growth hormones for height, and soon-to-come memory-enhancing drugs and anti-ageing drugs. I recently saw a report that even such an elementary intervention as providing extra oxygen to newborn babies can lead to significantly above average IQ scores when tested at eight years.[2]

But, more to the point, there has long been evidence from

natural history that Nature herself can intervene to boost performance levels if she so chooses—producing exceptionally well-endowed individuals all on her own.

These natural 'sports', if I may use that word, can take the form of individuals who grow up to have exceptional height, or strength, or beauty, or brains, or long life because they carry rare genes that bias them specifically in these directions. But, surprisingly enough, they can also show up as individuals who develop islands of extraordinary ability in the context of what we would more usually think of as retardation or pathology: epileptics with remarkable eidetic imagery, idiot savants possessed of extraordinary mnemonic faculties or musical talents, elderly patients with dementia who come out with superb artistic skills.[3]

Even enhanced beauty can come about as a secondary consequence of a developmental disorder. There is a syndrome called CAIS, complete androgen insensitivity syndrome, where male foetuses fail to respond to male sex hormones and consequently grow up to have the bodies of women: but not just any women—these transformed boys typically have enviable bodily proportions, long legs, unusual symmetry, glossy hair, pellucid skin (indeed, it is rumoured that several highly successful super-models and film actresses have been just such cases).

Now, if these examples mean what they seem to mean, we Darwinians perhaps have some explaining to do. For what the examples seem to suggest is that there *is* room for further progress in human evolution. In some respects anyway, human beings are not yet as highly evolved as in principle they could be (and perhaps they'd like to be). Perhaps Nature, after all, has *not* done the best that can be done for us—at least not yet. In which case the question we must ask is: Why?

Since there can be nothing wrong with the logic of the argument that says that any increase in a desirable trait will, when available, tend to go to fixation, the answer must be that the situation with regard to availability and/or desirability is not quite what it seems. That's to say, in these case we are interested in, either an increase in the trait in question, over and above what already typically exists, is actually *not an available* option within biological design space, or it is actually *not a desirable* option that would lead to increased fitness.

The first possibility, that the maximal level is actually not biologically attainable—or at any rate sustainable—is very much the answer that is currently in vogue among evolutionary biologists. In relation to IQ, for example, it is argued that while Nature has indeed done her best to design all human brains to maximize general intelligence, she is continually thwarted by the occurrence of deleterious mutations that upset the delicate wiring.[4] Or in relation to health and beauty, it is argued that while Nature has set us all up to have the best chance of having perfectly symmetrical bodies, pure complexions, and so on, there is no way she can provide complete protection against the ravages of parasites and other environmental insults during development.[5]

Yet, while there is surely something in this, it cannot be the whole story. For we have already seen some of the most telling evidence against the idea of there being this kind of upper ceiling: namely, that, despite the mutations, parasites, and other retarding factors that are undoubtedly at work, it *is* possible to intervene in quite simple ways—oxygen, foetal androgens, memory drugs, and so on—to enhance performance in particular respects; and, what's more, there do exist natural examples where, against the apparent odds, these problems have been overcome—those genetic variants, pathological compensations, and so on. In other words, it is clear that the reason why human beings typically do not reach these levels cannot be entirely that Nature's hands are tied.

The second possibility, that to reach for the maximum possible will not actually pay off in fitness, is in several cases both more plausible and more interesting.

This is the answer William Derham himself clearly preferred. What Derham pointed out is that even when a trait seems desirable, and indeed is so up to a certain point, in many cases it is possible to have too much of a good thing. Μηδὲν ἄγαν (*meden agan*), as the classical proverb goes: *do nothing in excess*. Too little and you will miss out on the benefits, but too much and you will find yourself incurring unexpected costs.

So, Derham argued, we should expect that true perfection must often lie in compromise. And in a perfect world, God— or as we now say, Nature—will have occasion to settle not for the maximum but for the 'golden mean'.

Thus, man's stature, for example, is not too small, but nor is it too large: too small and, as Derham put it, man would not be able to have dominion over all the other animals, but too large and he might become a tyrant even to his own kind. Man's physical countenance is neither too plain but nor is it too handsome: too plain and he would fail to attract the other sex, but too beautiful and he might become lost in self-admiration. Man's lifespan is neither too short, nor is it too long: too short and he would not have time to have enough children to propagate the species, too long and there would be severe overcrowding.

However, while this explanation seems to work nicely provided we choose our examples carefully, it is not clear it is going to work so well across the board. For there are other traits—intelligence, for instance—for which the dangers of excess are by no means so apparent and there might seem to be advantages to be had all down the line.

True, even when this is so, and advantage never actually turns to disadvantage, the returns to be had beyond a certain point might hardly be worth having. As Darwin himself noted: 'In many cases the continued development of a part, for instance, of the beak of a bird, or of the teeth of a mammal, would not aid the species in gaining its food, or for any other object.' Yet the fact remains that in some other cases—and intelligence may seem the prime example—the returns in terms of fitness would actually seem likely to remain quite high. Would there ever come a point where a human being, struggling for biological survival, would cease to benefit from being just that little bit *cleverer*, for instance? Darwin himself thought not: 'but with man we can see no definite limit to the continued development of the brain and mental abilities, so far as advantage is concerned.'[6]

Here, however, it seems that Derham was ahead of Darwin. Realizing that if a trait such as intelligence really were to be *unmitigatedly advantageous,* then God—or Nature—would have no excuse for settling for anything less than the maximum possible, and being under no illusion that human intelligence in general is in fact anywhere near this maximum point, Derham had no hesitation in concluding that increased intelligence must in reality be *disadvantageous*.

So, Derham reasoned, there must in fact be hidden costs to being too clever. What could they be? Well, Derham's idea was that, if man had been made any cleverer than he actually is, he would have been capable of *discovering things he ought not to know*. And, to prove his point, he proceeded to discuss three examples of discoveries to which man (at the time of writing, in 1711) had failed to find the key, and which it seemed obvious were beyond the powers of reasoning that God had given him. These are: in mechanics, the ability to fly; in mathematics, the ability to square the circle; and in navigation, the ability to judge longitude at sea.

Now, Derham admitted that he himself could not see what harm would come from man's being able to square the circle or judge longitude. But in the case of flying he had no doubt of the 'dangerous and fatal Consequence' that would follow if man were ever capable of taking to the skies:

As for Instance, By putting it in Man's Power to discover the Secrets of Nations and Families, more than is consistent with the Peace of the World, for Man to know; by giving ill Men greater Opportunities to do Mischief, which it would not lie in the Power of others to prevent; and by making Man less sociable, for upon every true or false Ground of Fear, or Discontent, and other Occasions, he would have been fluttering away to some other Place.[7]

We smile. But this idea is by no means entirely silly. The notion that it is possible for a person to be 'too clever by half' is one that has considerable folk credibility. And where there is folk credibility there is generally more than a little factual basis. Beginning with the story of Adam and Eve eating from the tree of knowledge, through Daedalus giving his son Icarus the wax wings with which he flies too close to the sun, to Frankenstein creating a monster he cannot control, myths and fairy tales offer us numerous examples of individuals who come to grief as a result of their being too clever or inquisitive for their own good. 'Curiosity killed the cat,' we say. 'More brains than sense.' And in the course of human history there must indeed have been many real life instances where human inventiveness has redounded in tragic ways on the inventor.

Not only in human history, but most likely in the history of other species, too. I am reminded of a report that appeared in the *British Medical Journal* some years ago:

Charles Darwin would doubtless have been upset had he known of the Coco de Mono tree of Venezuela. It apparently bears pods of such complexity that only the most dexterous of monkeys can open them and obtain the tasty almond-like nut. Once the nuts have been eaten the monkey's hair drops out and he soon expires—thus ensuring the survival of the least fit members of each generation.[8]

But note that the author is wrong to have written 'the survival of the *least fit*'; rather, he should have written, 'the survival of the *least skilled*'—for the lesson of this (possibly apocryphal) story is precisely that the least skilled may in fact be the most fit.

What is true for practical intelligence can surely be true for social intelligence as well. In an essay on the 'Social Function of Intellect', nearly thirty years ago, I myself raised just this possibility: arguing that Machiavellian intelligence, beyond a certain point, may turn against its owner because success in interpersonal politics becomes an obsession, leading him or her to neglect the basic business of productive living. 'There must surely come a point when the time required to resolve a "social argument" becomes insupportable.'[9]

The same surely goes for other capacities that we do not

usually think of as having a downside. I have no doubt a case could be made for the dangers of excess in relation beauty, say, or health. 'Too beautiful by half' and a person may run the risk of envious attacks by rivals. 'Too healthy by half' and . . . well, I'm sure there is *something* to be said against it.

So, let's call this line of explanation 'Derham's argument'. I think we can agree that Derham's argument is a perfectly reasonable argument. And in many cases it does provide a straightforward way of explaining why Nature has not pushed desirable traits to their biological limits.

But it is not the only possible way of explaining this apparent paradox. And it is not the one I am going to dwell on in this essay. For I think there is an even more interesting possibility out there waiting to be explored. It is an idea that was anticipated by another of those ingenious scientist-theologians at the turn of the seventeenth century, one Nehemiah Grew. And it is an idea that in some ways is the precise opposite of Derham's.

Derham's line was that too much of a good thing can get you into trouble. But Grew's line, expounded a few years earlier in his *Sacred Cosmology*, was that too much of a good thing can get you *out of trouble*, when actually it would be *better for you if you stayed in trouble*—better because trouble can be a blessing in disguise, *forcing you to cope by other means*.

Take the case of height and strength, for instance. Derham, as we have seen, suggested that God in his wisdom does not choose to set man's height greater than it is because if men were taller they might get into damaging, self-destructive fights. Grew, however, came up with the remarkable suggestion that God does not do so because if men were taller they might find life too easy, and consequently neglect to cultivate other essential skills.

Had the Species of Mankind been Gigantick . . . there would not have been the same Use and Discovery of his Reason; in that he would have done many Things by mere Strength, for which he is now put to invent innumerable Engines.[10]

Less strength because only comparative weaklings can be expected to invent innumerable engines! Let's call this 'Grew's argument'. It is a startling idea. It needs some unpacking. But then I think it may turn out to hold the key to several major puzzles about human evolution.

To see how Grew's argument can be developed, let's begin now from a more modern perspective (which is , as you may guess, where I myself set out from—having only later found my way back to Grew and Derham).

When the question is whether and how natural selection can arrive at the best design for an organism, a recurrent issue for evolutionary biologists is that of 'local maxima'.

To illustrate the problem, Figure 23 shows a typical 'adaptive landscape'. Here the biological fitness of a hypothetical organism, shown on the y-axis, is seen to vary *in an up-and-down way* as a function of some continuously varying phenotypic trait, shown on the x-axis. Under natural selection, which favours any small increase in fitness, there must be a

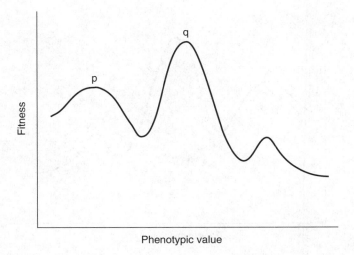

Fig. 23

Discoveries

tendency for the organism to evolve along the x-axis in whatever direction is upwards on the y-axis. Clearly in this case the
organism will be best off if it can in fact evolve to the absolute
maximum at point *q*. But suppose it is already at a local maximum at point *p*. Then, because it will have to go downwards
before it can continue upwards, it is stuck where it is.

Let's think of it in terms of the following analogy. Imagine
the graph with its local maxima is a ceiling with hollows, and
the organism is a hydrogen-filled balloon that is floating up
against it, as in Figure 24. The balloon would like to rise to the
highest level, but it cannot—it is trapped in one of those hollows.

This problem is, of course, typical of what happens to any
kind of system, evolving in a complex landscape, which seeks
to maximize its short-term gain and minimize its short-term
losses. There is no way such a system can take one step backwards for the sake of two steps forward; no way it can make a
tactical retreat so as to gain advantage later.

The situation is familiar enough in our own lives. Even we,

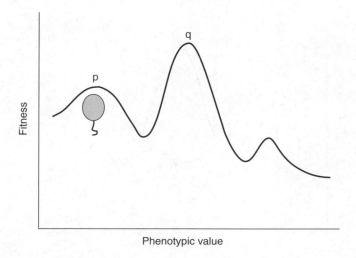

Fig. 24

who pride ourselves on our capacity for foresight, easily get trapped by the short-termism of our goals. 'The good', as is said, 'is the enemy of best': and, provided we are already doing moderately well, we are often reluctant to incur the temporary costs involved in moving on to something better. So, for example, we continue in an all-right job rather than enter the uncertain market for the quite-right one. We stick to techniques that work well enough rather than retrain ourselves in ways that could potentially work so much better. We stay with an adequate marriage rather than leave it for the distant prospect of a perfect one.

But let's look again at the balloon. Although it is true the balloon will never take the one step backwards for itself, it may still happen, of course, that it gets to have some kind of setback *imposed from outside*. Suppose a whirlwind blows through and dislodges it, or it gets yanked down by a snare, or it temporarily loses hydrogen. Then, once it has suffered this unlooked-for reverse, there is actually a fair chance it may float higher at its next attempt. In other words, there *is* a way the balloon can escape from the local hollow and achieve its true potential after all. But, oddly enough, what is needed is that something 'bad' will happen to it—bad in the short term, but liberating in the longer term.

And the same is true for us. Sometimes we, too, need to have a whirlwind blow through our lives before we will start over again and give ourselves the chance to move on to a new level.

Human history is full of examples of how seeming catastrophes can in fact be the catalyst for profitable change. People suffer dire poverty, or slavery, or are forced to migrate: they discover in their new world unprecedented riches. They have their cities razed and factories destroyed by bombs: they rebuild in ways far more efficient than the old. They lose their inherited wealth in a stock-market crash: they go to work to become healthier and wealthier than they ever were to start with.

Shakespeare, as always the superb student of human

nature, remarked how sweet are the uses of adversity 'which like the toad, ugly and venomous, wears yet a precious jewel in his head'.[11] Nietzsche wrote:

Examine the lives of the best and most fruitful men and peoples, and ask yourselves whether a tree, if it is to grow proudly into the sky, can do without bad weather and storms: whether unkindness and opposition from without . . . do not belong to the favouring circumstances without which a great increase in virtue is hardly possible.[12]

Even the children's film, *Chitty Chitty Bang Bang*, has a song that goes 'From the ashes of disaster grow the roses of success'.

But people are more interesting than balloons. The reason why disaster so often breeds success with human beings is not simply that it gives them, as it were, a new throw of the dice— so that with luck they may do better this time round (although it is true that luck may sometimes have a hand in it: many a human group forced to emigrate has by pure chance found a superior environment awaiting them abroad). The more surprising reason is that when people suffer losses and are obliged to find imaginative ways of replacing assets they previously took for granted, they frequently come up with solutions that bring a bonus over and above what they originally lost.

So, for example, when famine strikes, people who have previously foraged for themselves may discover ways of collaborating with others, which in the event bring in much more than the individuals could harvest even in good times on their own. Or, when they lose their vision from short sight, they may (they did!) invent spectacle lenses to make up for it, which in the event leads to the development of telescopes and microscopes and so provides them with better vision than they had before.

And it can happen on an individual level too. After Stephen Hawking (whom I knew as a childhood friend and who lived with my family for two years in the 1950s) suffered a debilitating neurological illness, he transformed himself from a relatively ordinary student into the extraordinary mathematical cosmologist he has since become. How did that happen? The novelist Martin Amis wrote recently: 'Hawking understood black holes because he could *stare* at them. Black holes mean oblivion. Mean death. And Hawking has been staring at death all his adult life.'[13] But the true reason is both more prosaic and more wonderful. Stephen Hawking, having lost the ability to write, could no longer work with algebraic formulae on paper, and was obliged to begin using geometric methods which he could picture in his mind's eye. But these geometric methods did more than substitute for the lost algebra; they gave Hawking ways of looking at things that his old algebra might never have revealed.

Remarkably enough, Albert Einstein told a similar story about how he himself gained from a disability. Einstein claimed that he was very late learning to speak and, even after he did, he found whole sentences tricky, rehearsing them in an undertone before speaking them out loud. This delayed development, Einstein said, meant that he went on asking childlike questions about the nature of space, time and light long after others had simply accepted the adult version of the world.[14]

Now, no one (at least no one who values his political correctness) would want to say that Stephen Hawking or the survivors of the Hiroshima bomb or the descendants of the African slaves were 'fortunate' to have had such a catastrophe in their personal or cultural background. Nonetheless, you can see how, whatever the subjects may feel at the time, in some cases it is objectively the case that what seems like ill fortune is actually good fortune.

So, we can come back to Nehemiah Grew. If ill fortune *in the short term* may indeed actually be good fortune *in the long term*, then it does make obvious sense that God himself will sometimes choose *to impose* ill fortune on his creatures in the short term *in order* that they achieve good fortune in the long term. That's to say, God may deliberately arrange to have human beings *born* less than perfect just in order that they find their way to *becoming* perfect. In particular, God may, as

Grew suggested, contrive to make human beings in certain respects weak and inadequate by nature, precisely because they will then be highly motivated to invent those 'innumerable engines'.

At any rate—God or not—here is the logic of these various examples:

- An accident or acquired defect threatens disaster by taking away a person's normal means of coping with a problem.
- The person is thereby given the incentive to find some alternative route to the same end.
- This alternative route, as it happens, not only overcomes the original problem but brings unanticipated benefits as well.

The burden of this essay is to argue that something very like this has played a significant part in biological evolution (and in particular the evolution of human beings):

- A mutation—a genetic accident—threatens to reduce an individual's fitness by removing some previously evolved means of solving a survival problem.
- The individual is thereby given the incentive to compensate by some novel behavioural strategy.
- This novel strategy, in the event, more than makes up for the potential loss in fitness and leaves the individual ahead.

I need hardly say that Nature—unlike God—cannot, of course, do things *deliberately*. She will never take a step backwards *in order* to take two steps forward. But I would argue that Nature may perhaps take a step backwards *by chance* in circumstances where the individual—with the help of his own intelligence, imagination, and culture—will be *more than likely* to take two (or at any rate more than one) steps forward.

And here I do mean *more* than likely. For I think there are theoretical grounds for supposing that, if and when an individual who finds himself deficient in some way is obliged to make up for his deficiency by replacing a genetically given strategy with an invented one, and succeeds, he will *more often than not* end up better off than if he has not had to do it. The simplest reason is this. Suppose we assume that in order to survive in competition with others the deficient individual has to come up with a new strategy that is *at least as good* as the genetically given one it is replacing. True, he may be able to get by with a new strategy that is *only just as good* as the original one, in which case he will have made one step backwards and only one step forwards, and will be back where he started. But it is a simple law of statistics that most of the possible strategies that are *at least as good* as the original will actually be *better* than it. (If you want to see why, consider, for example, why most of the people who pay at least as much tax as you do actually pay more tax than you do; why most of the prime numbers that are at least as high as 523 are higher than 523; and so on.) Hence the chances really are high that the deficient individual, if he survives at all, will adopt a strategy better than the one he started with.

Is it only human beings who will be able to get ahead in this surprising way? In principle all that is required is the capacity to replace genetically given features with invented ones. But in practice this probably does limit it—as a significant path for change—to our human ancestors. For there is, of course, one big barrier to its working out well even in the case of human beings: namely, the need for the individual who suffers the setback to be not only peculiarly inventive but peculiarly quick.

When Houdini was bound hand and foot and thrown into the lake, he could not afford to wait for his grandchildren to set him free. No more could one of our ancestors born with a biological deficiency leave it to later generations to make good what he had lost. The human brain, working within the context of human culture, is an organ—the *one* organ in nature?—that (quite unlike the genetically programmed body) *is* able to make astonishing progress within the span of an individual life.

Let's look, then, specifically to human prehistory. And let's look for scenarios to fit the logic spelled out above: where human ancestors can be seen as losing some genetically given beneficial trait (measured perhaps by comparison with their chimpanzee-like cousins, who still have it), therefore being obliged to make up for this loss by rapidly inventing a way round it (for which their cousins have no obvious need), and as a result moving unexpectedly ahead of the game (leaving those cousins standing—or extinct).[15]

I will offer two examples of losses and replacements in the course of human evolution to which this story may apply. First, the case of *the loss of body hair and the coming of fire-making*; second, the case of *the loss of memory capacity and the coming of abstract thinking*. The first is in some ways a 'toy example', which I shall present not because I am entirely serious about it, but because I think it nicely illustrates how the principle can work. But the second is a case about which I am serious to a degree.

Hair loss / Fire

Why have humans lost their body hair? Answers range from Desmond Morris's interesting suggestion in his book, *The Naked Ape*,[16] that hairlessness makes sexual intercourse more pleasurable and so promotes pair-bonding between human parents, to the standard theory that hairlessness reduces the dangers to human hunters of getting too hot when running after prey under the midday sun on the savannah.

These answers, within the conventional paradigm, seek to explain hairlessness as a simple direct benefit: human beings function better as lovers, say, or hunters without body hair than with it. Let's agree that such direct benefits, so far as they go, may be a factor (although this may not be very far).[17] But what about the much more obvious direct *costs*: the costs of getting cold?

The sun does not always shine, even in Africa. And while a hairless human being may benefit from not overheating when active at midday, the plain fact is the same human is bound to be at considerable risk of overcooling at other times of day, especially when inactive and at night. The dangers of cold are potentially severe. This is how the *Cambridge Encyclopedia of Human Evolution* summarizes the situation:

Although early human populations were established in quite cold climates long before the evolutionary appearance of *Homo sapiens*, modern humans have a very low tolerance of cold. Because we lack insulation such as fur and hair, nude exposure to still air temperature as high as 26°C causes constriction of blood vessels in the skin. At around 20°C, increased heat production, manifest as shivering, begins, and at 5°C inactive young adults may suffer such a reduction in brain temperature that they become unconscious in a few hours . . . Without the culture that produced clothing and fire and without access to some kind of shelter, our predecessors were limited to places where it never became colder than about 10°C.[18]

Unfortunately for those predecessors, however, there are few if any places in the whole world where it *never* becomes colder than 10°C. Even in much of central Africa, the minimum daily temperature regularly drops below 10°C at some season of the year. Nor has it been much different in mankind's ancestral past: in fact, during an Ice Age around 800,000 years ago, Africa must have been considerably colder than today.

We can picture those 'early human populations' that were 'established in quite cold climates long before the evolutionary appearance of *Homo sapiens*' as still having plenty of body hair and so as having no need as yet to go looking for a remedy. We know that at some stage between then and now, human beings did lose their hair, and they did indeed come through with the culture that produced clothing, fire, and shelter. The question is: what is the causal relationship here?

My hypothesis is that it was indeed hair loss that came first. That's to say, certain of those early humans—those who in fact became our ancestors—were *driven* to develop the arts of keeping warm precisely *because* they lacked sufficient hair and were becoming cold. But then, as things turned out, these individuals—and the trait for hairlessness—actually prospered: for the fact is that *the cultural innovations brought a significant, though unanticipated, premium.*

Consider in particular the case of fire. No doubt people first learned to make and tend fires for no other reason than to stave off cold. Yet we now know—as the first fire-makers presumably did not—that once the fires were burning, a host of other benefits would soon become available. For the very same fire that provided warmth could help keep predators away; it could provide light; it could be used to harden stone and wooden tools; it could be used for cooking food, rendering meat and vegetables more digestible and killing parasites; and—perhaps the biggest if least tangible benefit of all—it could provide a focus for family and friends to gather round, to exchange gossip and information and cement social bonds.

The upshot is that the biological setback of hair loss—in so far as it was a precondition for the cultural innovations— would have actually brought a net gain in biological fitness. Hairlessness, therefore, would have proved to be on balance an evolutionarily adaptive trait, and so would have been set to spread through the population at a genetic level.

But why, you may ask, should it be true that hairlessness was a *precondition* for fire-making? If fires could bring all those added benefits besides warmth, why did not early human beings hit on the idea of making fires anyway—even before they lost their hair?

The probable answer is that these other benefits simply did not provide the right kind of *psychological incentive.* Human beings, if and when they are cold, have an instinctive liking for the warmth of fire. But they have no comparable *instinctive* liking for parasite-free cooked meat, or fire-hardened knives, or predator-scaring flames, or even camp-fire-facilitated social gatherings. Even if these other benefits would have come to be appreciated in good time, their absence would have been unlikely to provide the necessary shock to the system that was required to get fire invention going.

I think the archaeological record supports this reading of

events. The first evidence of man-made hearths is at sites dating to about 400,000 years ago. However, for reasons that have been thought to be something of a mystery, fire-making does not seem to have caught on, and hearths remain remarkably rare in areas known to have been lived in by humans—until about 150,000 years ago, after which they soon become much more common.

The archaeologist Ian Tattersall comments: 'The advantages of domesticating fire are so great that, once technologies for its control had been developed, it's hard to see why they should not have spread rapidly.'[19] I'd say the reason may well have been that it all depended on the history of hair loss—and people *first* feeling cold.

Memory loss / Abstraction

Have human beings lost their memories? The fact that modern human beings have less hair than their chimpanzee-like ancestors is obvious and indisputable. But that they have less memory capacity than their ancestors is not something generally acknowledged.

My chief reason for claiming it is so is some little-known research on 'picture memory' in chimpanzees, undertaken by Donald Farrer nearly forty years ago.[20]

In his experiment, Farrer gave three chimpanzees a 'match-to-sample task', where the subject was presented with an array of pictures as shown below, and was rewarded for finding and touching the picture in the bottom row that matched the sample picture at the top.

Twenty-four different combinations were presented, as shown in Figure 25. But these were not a random selection. In fact, the same line-up in the bottom row never recurred with a

Problem No.	Pictures				Correct position
	Lever No.1	Lever No.2	Lever No.3	Lever No.4	
1	+	(G)	|	×	4
2	△	—	□	◎	2
3	(G)	(W)	+	×	3
4	(B)	—	(R)	□	1
5	(G)	△	—	×	2
6	+	×	(R)	(B)	4
7	|	(G)	(W)	△	3
8	|	(G)	—	◎	1
9	+	◎	(OR)	×	3
10	△	□	|	×	2
11	(W)	◎	(B)	|	4
12	◎	(OR)	□	—	1
13	(OR)	(W)	—	(G)	1
14	×	□	(OR)	(R)	2
15	(R)	—	+	□	3
16	△	(G)	|	(R)	4
17	×	□	◎	(B)	3
18	(B)	|	×	—	4
19	(B)	+	□	—	1
20	+	(G)	|	(B)	2
21	×	(W)	—	(B)	2
22	△	×	(G)	(R)	1
23	□	(B)	×	|	3
24	△	—	□	(R)	4

Fig. 25. The twenty-four problems in the picture memory test

different matching picture above: so that, for each particular line-up in the bottom row, there was only ever one correct answer. This meant that, to perform well on the task, the subject did not actually have to learn the 'match-to-sample' rule at all, but could instead learn by trial and error which one picture was correct for each of the twenty-four possible line-ups.

Farrer trained his chimpanzees until they were getting 90 per cent correct. And then, so as to find out which strategy they had in fact used, he gave them a series of test trials in which the bottom row of pictures was presented *without the sample on top being visible*—so there was now no way of applying the rule, and only rote memory for each particular line-up could possibly suffice. Astonishingly, the chimps continued to perform as well as before, selecting the 'correct' picture between 90 per cent and 100 per cent of the time. Clearly they had in fact learned the task by rote. Farrer's conclusion was that chimpanzees do indeed have a supra-human capacity for 'picture memory'—or 'photographic memory'.[21]

But if chimpanzees have this capacity today, then, unless they have acquired it relatively recently, it is a fair guess that our human ancestors also had it to begin with. And, if our ancestors had it, then modern humans must indeed have *lost* it.

Why? Why lose a capacity for memorizing pictures, when prima facie there can only be immediate costs—just as there are to losing hair? I suggest the reasons for memory loss were indeed structurally of the same kind as the reasons for hair loss: when human beings lost their memories they were obliged to solve their problems some way else, and this some way else turned out to be hugely advantageous.

And Farrer's experiment provides an immediate lead to what the advantage of this some way else might be. If you or I were given the same match-to-sample task, we with our poor memories would find it nearly impossible to solve it by rote learning and so we would do the modern human thing and search for some higher-order pattern in the data. But, lo and behold, once we identified this pattern, the match-to-sample *rule*, we would have not merely caught up with the chimpanzees on this one task, we would have inadvertently qualified ourselves to solve a whole range of problems *we have never met before.*

Suppose, for example, that after being trained with the original set of twenty-four combinations of pictures, we were now given a combination that was not part of the original set, such as the one below, where the same line-up below appeared with a different picture above:

The chimpanzee in this new situation would presumably continue to choose the ✩; but we, knowing the rule, would choose the ♣.

In short, the use of rules—and abstraction in general—allows knowledge of the world acquired in one situation to be applied in another: a capacity that is the very ground of our human cleverness and creativity.

There is increasing evidence that chimpanzees are in fact surprisingly backward when it comes to rule learning and abstract thinking. It is not that they are completely incapable of it—chimpanzees certainly *can* learn the match-to-sample rule when there is no other way—but, rather, that in many situations they simply do not bother with it. The work of Daniel Povinelli, in particular, has demonstrated how superficial chimpanzees' understanding generally tends to be: how they seldom, if ever, classify things on the basis of essential properties or interpret events in terms of hidden causes. And Povinelli himself raises the possibility that this failure to seek explanations beneath the surface of appearances results from the fact that, for chimpanzees, appearances are indeed too memorable—and salient—for them to let go.[22]

But even better evidence of what might have become of us if we had retained our memories is provided by the rare cases of modern human beings who, for whatever reasons, still do possess a chimpanzee-like capacity. And most remarkable is the case of the Russian, Mr S., described by Alexander Luria in his book *The Mind of a Mnemonist*.

In the 1920s, S. was a young newspaper reporter in Moscow who, one day, got into trouble with his editor for not taking notes at a briefing. By way of excuse, he claimed he had no need for notes since he could remember everything that had been said, word for word. When put to the test, he soon demonstrated that he was in truth able to recall just about every detail of sight and sound he had ever encountered.

For the rest of his life S. was intensively investigated. In laboratory tests he was shown tables of hundreds of random numerals, and after looking at them for just a few minutes, he was able to 'read off from memory' exactly what was there—forwards, backwards, diagonally, or in any way requested. What is more, after years of memorizing thousands of such tables, he could go back to any particular one of them on any particular date and recollect it perfectly, whether it was an hour after he first saw it or twenty years. There really did seem to be almost no limit to his memory capacity.

Yet S., too, was surprisingly backward. He remembered everything but he understood next to nothing. The simplest conceptual structures passed him by. He completely failed to grasp the connectedness of things. For example, when he was given a list of words, some of which were names of birds, he could remember the whole list, but he simply could not pick out the birds as a separate group. When given a chart containing the following series of numbers to remember:

$$1 \ 2 \ 3 \ 4$$
$$2 \ 3 \ 4 \ 5$$
$$3 \ 4 \ 5 \ 6$$
$$4 \ 5 \ 6 \ 7$$

etc.

he proceeded to recall the entire series, unaware that the numbers progressed in a simple logical order. As he later remarked to Luria: 'If I had been given the letters of the alphabet arranged in a similar order, I wouldn't have noticed their arrangement.'[23]

It should be said that, as with the chimpanzees, S.'s problem was almost certainly not that he was entirely incapable of abstract thinking; it was just that he had little if any inclination for it. Memorizing was so comparatively easy for him that he found abstract thinking unnecessary and uninviting.

So, S.'s plight perfectly illustrates what is at stake.[24] In fact I'd suggest S. can be regarded (with due respect) as having been a living exemplar of that earlier stage in human evolution when our ancestors all had similar qualities of mind: similar strengths in the memory department and consequently similar weaknesses in understanding. There but for the grace of evolution, go you and I.

What happened, however, was that memory loss liberated us. Those of our ancestors unfortunate enough (but fortunate) to suffer a sudden decline in memory capacity had to discover some way of making up for it. And the happy result was that they found themselves reaping a range of unanticipated benefits: the benefits that flow from a wholly new way of thinking about the world. No longer *able* to picture the world as made up of countless particular objects in particular relationships to each other, they *had* to begin to conceive of it in terms of categories related by rules and laws. And in doing so they must have gained new powers of predicting and controlling their environment.

In fact, there would have been additional ways in which human beings, once their memories begin to fail, would have tried to make up for it. No doubt, for example, they would soon enough have been taking measures, just as we do today, to organize their home environment along tidy and predictable lines; they would have been making use of external ways of recording and preserving information (the equivalent of S.'s absent notebook!); they would have been finding ways of sharing the burden of memorizing with other human beings. And all these tricks of off-loading memory into the 'extended mind' would certainly have increased the net gain.

But, more significant still, by taking these various steps to compensate for their poor memory, our ancestors would have inadvertently created the conditions required for the development of *language*. Quite why and when human language took off remains a scientific mystery. But, before it could happen, there's no question several favouring factors had to be in place: (i) human beings must have had minds prepared for using high-level concepts and rules, (ii) they must have had a cultural environment prepared for the externalization of symbols, and (iii) they must have had a social structure where individuals were prepared for the sharing of ideas and information. Each of these factors might well have arisen, separately, as a way of compensating for the loss of memory.

Once these several factors *were* in place, things would most likely have developed rapidly. Not only would language have proved of great survival value in its own right, but there could have been an emergent influence moving things along. I have been emphasizing how loss of memory would have encouraged the development of language, with the causal influence running just in one direction. But the fact is that memory and language can interact in both directions. And there is reason to believe that in certain circumstances the use of language may actually weaken memory: as if at some level linguistic descriptions and picture memory are rivals—even as if words actively erase pictures from memory.

Consider the following experimental finding. People are asked to remember a visual scene, under two conditions: those in one group are asked to describe the scene in words, while those in a control group do not describe it. Later, both groups are given a recognition test in which they have to say whether particular details—including some which were not in the description—were present in the original scene. It turns out that those who have described the scene in words are likely to have retained less of the information about details than the control group.[25]

But, now, think about what this might mean for the early stages of language evolution. If the effect of using language was indeed to undermine memory, while the effect of memory being undermined was to promote the use of language, there would then have been the potential for a virtuous circle, a snowball effect—with every advance in the use of language creating conditions such as to make further advances more probable. The language 'meme' (compare a software virus) would effectively have been manipulating the environment of human minds so as to make its own spread ever more likely.

Thus, I'd say it really could have been the same story as with hair: the biological setback of memory loss—in so far as it was a precondition for the mental innovations—brought a net gain in biological fitness. Memory loss proved on balance to be an evolutionarily adaptive trait, and so was set to spread through the population at a genetic level.

But again you may ask: why should memory loss have been a *precondition* for these innovations? If abstract thinking is so beneficial, why would people not have adopted it anyway, irrespective of whether their memories had let them down?

The examples of S. and the chimpanzees do seem to confirm that, for so long as memory remains too good, there really is not sufficient immediate payoff to do anything else (even if the capacity is latent). And yet the extent of the mental laziness is certainly surprising. One of the founding fathers of cognitive psychology, Frederick Bartlett, made much of what he called 'the effort after meaning', which he supposed to be an instinctive delight that all human beings take in making sense of things: and it does seem strange that this *instinct*—if it is such—should have been so little evident in a case like S. Possibly the answer is that it kicks in only after a first attempt at gaining cognitive control at a lower level by rote memory has failed. Maybe S., like a man who has never been thirsty and so never known the joys of slaking his thirst at a cool spring, simply never had cause to *feel* as we do about rules.

Let me turn to what we can discover from the archaeological record. The evidence for when human beings first responded to memory loss by adopting new styles of thinking is never going to be as clear as the evidence for when they responded to hair loss by making fires. However, there could be indirect evidence in the form of artefacts or traces of cultural practices that show the clear signature of human beings who either *were* or *were not* thinking in particular ways. Archaeologists do now claim, on the basis of traditions of stone tool making, in particular, that humans were using abstract concepts—and possibly verbal labels—as long ago as half a million years.[26] In which case, presumably, it would follow that the crucial loss of memory capacity must have occurred before that time.

I will not presume to criticize this interpretation of the stone tool evidence. But I shall, nonetheless, refer to an observation of my own which, if it means what I have elsewhere suggested it means, tells a very different story. This is the observation (described at length in Chapter 12) of the uncanny resemblance between Ice Age cave paintings of 30,000 years ago, and the drawings of an autistic savant child, Nadia, living in the 1960s: a child with photographic memory but few, if any, mental concepts and no language.

We have only to compare reproductions of the cave paintings side by side with the child's to be struck by how similar they are in style, in content and in execution (see Figures 8–22). But the reason this resemblance is so surprising and important is that this level of graphic skill is never found in modern-day children in whom thinking and language have developed normally. Indeed, there are good grounds for believing that Nadia (and the few other savant children like her) could draw as she did only *because* she pictured the world in a non-conceptual way.

I have argued that there is a real possibility that the cave artists themselves had savant-like minds, with superior memories and undeveloped powers of abstract thinking. In that case the loss of memory capacity and the development of modern styles of abstract thinking might in fact have come remarkably recently, only a few tens of thousands of years ago.[27]

Let this be as it may. You do not need to be convinced by every detail of the story to accept we are on to an interesting set of possibilities here. Let's ask what else falls into place if it is right.

An obvious question is: if there have been these steps backwards in the design of minds and bodies in the course of human evolution, just how could they have been brought about genetically? For, in principle, there would seem to be two very different ways by which a genetically controlled feature, such as hair or memory, could be got rid of, if and when it was no longer wanted: it could be *removed,* or it could be *switched off.*

It might seem at first that removal would be bound to be the easier and more efficient option. But this is likely to be wrong. For the fact is that in order to remove an existing feature, as it appears in the adult organism, it would often be necessary to tamper with the genetic instructions for the early stages of development, and this might have unpredictable side effects elsewhere in the system. So in many cases the safer and easier course would actually be to switch the feature off—perhaps by leaving the original instructions intact and simply inserting a 'stop code' preventing these being followed through at the final stage.[28]

The most dramatic evidence for switching off rather than removal of genetic programs is the occurrence of so-called 'atavisms'—when ghosts of a long-past stage of evolution re-emerge as it were from the dead. To give just one remarkable example: in 1919, a humpback whale was caught off the coast of Vancouver which at the back of its body had what were unmistakably two hind legs. The explanation has to be that, when the hind legs of the whale's ancestors were no longer adaptive, natural selection eliminated them by *turning off* hind-leg formation, while the program for hind legs nonetheless remained latent in the whale's DNA—ready to be reactivated by some new mutation that *undid the turning off.*

Do such atavisms occur in the areas of human biology we are interested in?

Let's consider first the case of hair. If body hair has been turned off, does it ever happen that it gets turned on again? The answer is most probably: Yes. Every so often people do in fact grow to have hair covering their whole bodies, including their faces. The best-documented cases have occurred in Mexico, where a mutant gene for hairiness (or, as I am suggesting, the *return* to hairiness) has become well established in certain localities.[29]

But how about the case of picture memory? We have seen two remarkable cases where the capacity for perfect recall popped up, as it were, from nowhere: the mnemonist S. and the idiot savant artist Nadia. But lesser examples of much better-than-average memory do turn up regularly, if rarely, in a variety of other situations. The capacity for picture memory is actually not uncommon in young children, although it seldom lasts beyond the age of five years. In adults it sometimes occurs as an accompaniment to epilepsy, or certain other forms of brain pathology, and it can emerge, in particular, in cases of senile dementia associated with degeneration of the fronto-temporal areas of the cortex. Several cases have recently been described of senile patients who, as they have begun to lose their minds in other ways, have developed a quite extraordinary—and novel—ability to make life-drawings.[30]

There is also evidence, albeit controversial, that certain racial groups possess exceptional memory. The anthropologist Eugene Marais described in his book, *The Soul of the Ape*, a variety of tests he undertook with Kalahari Bushmen from which he concluded that the Bushmen have preserved a capacity for memorizing features of the landscape that civilized human beings have subsequently lost.[31] More recently, and more reliably, studies of Australian Aborigines show that they typically perform brilliantly—and much better than European whites—at 'Kim's game'-type memory tasks, where they are asked to look at a complex spatial array of objects and then, after the array has been disassembled, to put everything back in its right place.[32]

In none of these cases can we be sure where the exceptional ability is coming from. But, assuming that no special training has taken place, it does seem likely that what is occurring is some kind of release effect. That is to say, the extra memory capacity is emerging spontaneously as and when some influence that normally keeps it in check is lifted.[33]

Even in S.'s case it can be argued several ways. Nonetheless, I dare say S. was indeed an example of the kind of atavism that our theory of evolved memory loss predicts might sometimes occur: a man born to remember because of a congenital absence of the *active inhibition* that in most modern human beings creates forgetting.

Luria wrote: 'There is no question that S.'s exceptional memory was an innate characteristic, an element of his individuality.' Interestingly enough, the trait seems to have run in S.'s family, with both his parents and a nephew also having unusually good memories.[34]

Was there a simple genetic cause involved here? We do not know. But, let's suppose for a moment it *was* so. Then perhaps I may be allowed one further speculation. If such a trait can run in a family, then presumably it could run in a whole racial group. In which case the superior memory capacity of Australian Aborigines (assuming the claims for this stand up) may in fact be evidence that the Aborigines as a group are carrying genes for a (newly acquired?) atavism.

This ends my main case. I have dwelt on the examples of hair loss and memory because I reckon they provide textbook examples of how the Grew effect might work. However, I realize these are not the examples that got the discussion going at the beginning of this essay. And I owe it to you, before I end, to try these ideas in relation to beauty loss and intelligence loss. These cases are too complicated to treat here with due seriousness. But I will try not to disappoint you entirely.

Beauty

Why are most people less than perfect beauties—certainly not as beautiful as they would like to be, and probably not as beautiful as they could be if only they were to have the right start in life (the right genes, the right hormones)?

I have mentioned already how one way of explaining the general mediocrity is to attribute it to the sheer difficulty of achieving a perfect face and body when everyone is up against the mutations, parasites, and accidents that dog all human development. Beauty has evolved as a form of sexual display, designed by sexual selection to show off each person's qualities as a mate: and this means—to make matters worse—that the dimensions of beauty most closely watched by other people are likely to be precisely those on which it is most difficult to do well. Thus, we particularly admire bodily symmetry, high cheekbones, unblemished complexion, and so on, precisely because not everyone, not even most, can achieve a perfect score in these respects.[35]

Let's allow that this is partly it. Yet I make bold to assert that most people's scores are so far off perfect—in fact so much closer to frank plainness than to beauty—that something else must be going on. Perhaps this something else is indeed the *active masking* of beauty.

Yet why should beauty be masked? Derham's view, as we saw, would be that too great beauty (like too great anything) can get a person into difficulties, perhaps by making him or her unduly narcissistic or the object of too much attention from the opposite sex. But Grew's view, I assume, would be that the problem with too great beauty is not simply that of attracting too much attention, *but of attracting it too easily*, and so having no incentive to compete for mates by other means. By contrast, those individuals blessed with lack of beauty will have a strong incentive to make up for it. These relatively plain men and women will indeed have to have recourse to the 'innumerable engines' by which individuals can and do try to compensate for their deficits in physical attractiveness—by being particularly kind, witty, artistic, creative, charming, houseproud, and so on. But in so doing they will more than make up in attractiveness and fecundity for what they lacked to start with.

My sister Charlotte will not mind me telling a story about her, that she once told me. Charlotte remembers that, as a teenage girl, she consulted a mirror and decided she was never going to win in love or in work by virtue of her looks. So she came to a decision: instead of competing on the unfair playing field that Nature had laid out for her, she would systematically set about winning by other means. If she could not be especially beautiful, she would make herself especially *nice*. And so, in fact, she did—going on to lead an enviably successful life all round.

There are other grand examples. George Eliot had such a poor opinion of her chances in the conventional marriage market that she took to writing novels instead. And, on the man's side, Tolstoy complained that he could 'see no hope for happiness on earth for a man with such a wide nose, such thick lips, and such tiny grey eyes as mine', and so he too decided he might as well make the best of a bad job and develop himself as a writer.

Perhaps this has been a recurring theme in relatively recent human history. But I emphasize *relatively recent* because I guess this route to success for those who lack beauty would have opened up only once human beings had already evolved the minds and the culture to enable them to seize the opportunity to develop and showcase their compensatory talents. We can imagine, then, an earlier stage of human evolution when physical beauty was actually more important—and so presumably more prevalent—than it is today: because, frankly, beauty was *it* and there were few if any ways of competing at a cultural level.

Intelligence

And why are most people so far off being highly intelligent? Given that the human brain is capable of creating a Newton or a Milton, the fact that the average person is—well—only so averagely good at the kinds of reasoning and imaginative tasks that form the basis for intelligence tests is, to say the least, regrettable.

Again there is a possible explanation in terms of mutation load and depredations to the brain during development. And again it is possible that these effects are amplified by sexual selection. The evolutionary psychologist Geoffrey Miller argues persuasively that a person's performance on an IQ test (or its equivalent in courtship rituals) can function, like beauty, as a way of displaying his or her genetic and physiological health, so that the signs of intelligence and creativity we most value are likely to be precisely those which only the highest-quality individuals can reliably lay on.[36]

Let's allow, once more, that there is something in this. But I seriously question whether it is the whole explanation. An IQ of 100 is not just a notch or two below the best, it would seem to be in a completely different league. And with half the human population scoring less than this, we should be thinking again about the possibility of active masking.

Yet, why mask intelligence? Derham's view would be that individuals with too great intelligence may be at risk of bringing destruction on themselves by finding answers to problems that are better left unsolved. But Grew would point to a very different risk: not that there is a cost to getting to the solution as such, but that there may be a cost to getting to it *effortlessly and unaided*. By contrast, those individuals with a relative lack of brain power will be bound to resort to more roundabout means, including again a variety of those cultural 'engines'. They will spend more time over each problem, make more mistakes, use more props—all of which may bring serendipitous rewards.[37] But, most important, I would say, they will be obliged to seek assistance from other people—and so will gain all that flows from this in terms of camaraderie and social bonding.

Perhaps you remember from your schooldays the risks of being the member of the class whose mind works too fast, the know-all who races ahead on his own while others seek help from the teacher and each other. These days kids have a new word for it: to be too bright in class is distinctly not 'cool'. But it is not only schoolkids who will ostracize an individual who may be too self-sufficient to work well with the team. I heard the following CNN news report on the radio recently:

The police department in New London, Connecticut, has turned down a man as a potential member of the force because he has too high an IQ. A spokesman for the recruiting agency says: 'The ideal police recruit has an IQ of between 95 and 115.' The man rejected had an IQ of 125.[38]

An IQ of 125 is about the level of most college students. Not, you might think, so exceptionally, dangerously intelligent. Nonetheless, I suspect the recruiting agency—and our classmates, and in the long run natural selection, too—are better judges than we might like to believe of what plays out to people's best advantage. When individuals are so clever that they have no need to work with others, they will indeed tend to shift for themselves, and so lose out on the irreplaceable benefits of teamwork and cooperation.

'But many that are first shall be last; and the last shall be first.'[39] It is hard for a rich man to enter into the kingdom of God. It is hard for a furry ape to catch on to making fires, for a mnemonist to become an abstract thinker, for a beautiful woman to become a professor, for a man with a high IQ to get into the New London police force. The warm-coated, the memorious, the beautiful, the smart, shall be last; the naked, the amnesic, the plain, the dull shall be first.

Many of us must find the teaching of Jesus on this matter—and the parallel conclusions we have come to in this essay—paradoxical. I think these conclusions are and will probably always remain deeply and interestingly counter-intuitive.

What we have discovered is that what people consider *desirable* is in fact often less than *optimal*. But, then, the question remains: why do they still desire it? Why has Nature saddled human beings with a yearning to possess qualities they are better off without?

The answer, I suppose, is that this is part of the deal. If human beings did not retain the *ambition* to regain the capacities they have lost—if they did not *envy* those who are by nature warmer, better informed, more sexually attractive, more brilliant than they—they would not try sufficiently hard to compensate for their own perceived deficiencies by alternative means. They have to be continually teased by the *contrast* between what they are and what they imagine they might be before they will go on to on to take the two steps forward that secure their place in history. It is the individual who, harking back to the time when people were angels, still has a vision of what it must be to fly like a bird, who *will* eventually learn how to take to the skies (and so prove old Derham's worst fears wrong).

Lord Byron knew this, and wrote a poem about it in 1824, *The Deformed Transformed* —a poem that captures in eight lines all that this chapter is about.

> Deformity is daring.
> It is its essence to o'ertake mankind
> By heart and soul, and make itself the equal—
> Aye, the superior of the rest. There is
> A spur in its halt movements, to become
> All that the others cannot, in such things
> As still are free to both, to compensate
> For stepdame Nature's avarice.[40]

PRETENCES

15

Tall Stories from Little Acorns Grow

My great grandmother began a book about her family as follows: 'Writing in the middle of the twentieth century, it is something to be able to say that I remember clearly my mother's father, born in 1797.'[1] She was a story-teller, and so in turn am I. Other people might have danced with a man who had danced with a girl who had danced with the Prince of Wales. But I, as a boy, had sat in the lap of someone who had sat in the lap of someone who had fought alongside Wellington at Waterloo.

Waterloo? Well, I had done the sums, and there was no question the basic facts were on my side. Born in 1797, my three-greats-grandfather must have been eighteen at the time of the big battle: he *could* have been there—and no doubt *would* have been there, if he had not (by unfortunate mischance) become a Nonconformist minister instead. At any rate, the Waterloo version was how I told it to myself and to my friends at school. I was not going to spoil a good story for a ha'penny worth of embellishment.

I would like to be able to say that I have now grown up. I have, however, listened to myself over the years and, in the face of the evidence that continues to accumulate, I will not pretend I am so much more reliable now than I was then. Indeed—I say this simply as a piece of scientific observation— I find myself embarrassingly often being overtaken by my tongue. I don't mean just in relation to people I have known, or things I've heard, but in relation to things I claim that I myself have seen.

For example (I should not tell you this, but in the interests of objectivity I will), since returning from Moscow at the beginning of June this year [1987], I have caught myself several times recounting to attentive friends the story of how I saw that plucky German pilot land his Cessna aeroplane on the cobbles beside the Kremlin wall. In one version I saw it from my hotel window; in another I was actually standing in Red Square.

Now it so happens I *was* in Red Square on the afternoon the plane came in; I did see it later that evening. But, as luck would have it, I was not actually there when it arrived. In my first accounts I told the literal truth. But as the story got repeated, so it got embellished—to the point where I now have to remind myself that it's not true.

I mention all this, not in a spirit of confession, but rather as a contribution to the study of human nature. For I do not think that it is only me. Not everyone, admittedly, owns up to telling lies. But there are surely enough of us about who enough of the time go in for a bit of innocent exaggeration, that we ought to raise the general question: why?

The obvious answer is that it is a way of gaining unwarranted respect. One of the basic principles of social life is that society accords a special place to the person *who has something to say*. To have been present at, or even intimately linked to, a newsworthy event makes one a valuable resource to the community. One becomes the person 'who was there', who has information to impart. It gives one what might be called 'witness power'. No wonder, then, that we are sometimes tempted to assert that we have been present at events where we were not.

Yet this answer would cover a range of sins, over and above those that I, at least, would own up to committing. The truth is I seldom if ever *invent* stories, I merely *improve* on them. And psychologically—perhaps morally as well—there seems to be a world of difference between concocting a story for which there is no basis whatsoever and merely taking a little licence with existing facts. Nothing will come of nothing. In no way would I have claimed that I saw the Cessna landing, *if I had not seen it at all*. And, even with that ancestor who fought at Waterloo, I would not have told the story *if I had not had at least a half-truth on my side.*

I am not saying I never make things up from scratch. But as a rule I am not tempted unless and until the real world provides some kind of quasi-legitimate excuse. I must have, as it were, a 'near- miss' in my experience—perhaps I was *nearly* there, or something happened to the next person along, or the fish got off the hook, or the lottery ticket was just one number wrong. And then I may not be able to control it.

In real life, a miss may be as good as a mile. But fantasy works differently. These near-misses provide lift-off for imagination, a ticket to the world of might-have-beens. 'The reason', W. H. Auden wrote, 'why it is so difficult for a poet not to tell lies is that in poetry all facts and all beliefs cease to be true or false and become interesting possibilities.'[2] The poet in us is evidently responsible for some of our more interesting prose.

16

Behold the Man: Human Nature and Supernatural Belief

When you keep putting questions to Nature and Nature keeps saying 'No', it is not unreasonable to suppose that somewhere among the things you believe there is something that isn't true.

It might have been Bertrand Russell who said it. But it was the philosopher Jerry Fodor.[1] And he might have been talking about research into the paranormal. But he was talking about psycholinguistics. Still, this is advice that I think might very well be pinned up over the door of every parapsychology laboratory in the land, and (since I may as well identify both of my targets on this occasion) every department of theology too. It will serve as a text for this essay, alongside the following that *is* from Russell:

I wish to propose for the reader's favourable consideration a doctrine which may, I fear, appear wildly paradoxical and subversive. The doctrine in question is this: that it is undesirable to believe in a proposition when there is no ground whatever for supposing it true.[2]

A few years ago I had the good fortune to be offered a rather attractive fellowship in Cambridge: a newly established research fellowship, where—I was led to understand—I would be allowed to do more or less whatever I wanted. But there was a catch.

The money for this fellowship was coming from the Perrott and Warrick Fund, administered by Trinity College. Mr Perrott and Mr Warrick, I soon discovered, were two members of the British Society for Psychical Research who in the 1930s had set up a fund at Trinity with somewhat peculiar terms of reference. Specifically the fund was meant to promote 'the investigation of mental or physical phenomena which seem to suggest (a) the existence of supernormal powers of cognition or action in human beings in their present life, or (b) the persistence of the human mind after bodily death'.[3]

Now, the trustees of the fund had been trying, for sixty years, to find worthy causes to which to give this money. They had grudgingly given out small grants here and there. But they could find hardly a single project they thought academically respectable. Indeed, it sometimes seemed that the very fact that anyone applied for a grant in this area was enough to disqualify them from being given it. Meanwhile the fund with its accruing interest grew larger and larger, swelling from an initial £50,000 to well over £1 million. Something had to be done. Eventually the decision was made to pay for a senior research fellowship at Darwin College (not at Trinity) in the general area of parapsychology and philosophy of mind, without any specific limitations. The job was advertised. I was approached by friends on the committee who knew of my outspoken scepticism about the paranormal. And—to cut a long story short—in what was something of a stitch-up I was told the job was mine on the understanding that I would do something sensible and not besmirch the good name of the college by dabbling in the occult or entertaining 'spooks and ectoplasm'.

Things do not always work out as expected. You know the story of Thomas à Beckett—King Henry II's friend and drinking companion, whom he unwisely appointed as Archbishop of Canterbury on the understanding that he would keep the church under control? Thomas, as soon as he had the job, did an about-turn and became the church's champion *against* the king. I won't say that is quite what happened to me. But after my appointment I too underwent something of a change of heart. I decided I should take my commission as Perrott–Warrick Fellow seriously. Even if I could not believe in any of this stuff myself, I could at least make an honest job of asking about the sources of other people's beliefs.

So I set out to see what happens when people put a particular set of 'questions to Nature about the supernatural'. The questions Messrs Perrott and Warrick would presumably have wanted to have had answers to would be such as these: '*Do* human beings have supernormal powers of cognition or action in their present life?' Can they, for example, communicate by telepathy, predict how dice are going to fall, bend a spoon merely by wishing it? '*Does* the human mind persist after bodily death?' Can the mind, for example, re-enter another body, pass on secrets via a medium, reveal the location of ancient buried treasure? And so on.

The trouble is, as we all already know, that when you ask *straight* questions like this, then the *straight* answer Nature keeps on giving back is indeed an uncompromising 'No'. No, human beings simply do not have these supernormal powers of cognition or action. Carry out the straightforward experiments to test for it, and you find the results are consistently negative. And no, the human mind simply does not persist after bodily death. Investigate the claims, and you find there is nothing to them. It turns out there really are 'laws of Nature', that will not allow certain things to happen. And these natural laws are not like human laws which are typically riddled with *exceptions*: with Nature there are no bank holidays or one-off amnesties when the laws are suspended, nor are there any of those special people, like the Queen of England, who are entitled to live above the laws.

This is a shame, perhaps. But there we are. As Ernest Hemingway once said, 'It would be pretty to think otherwise'. But it is not otherwise. This is the world we live in. And so, you might suppose, most people would long ago have grudgingly accepted that some—or indeed rather a lot—of the things they would obviously have liked to believe in are not true. ESP, psychokinesis, trance channelling, and so on really do not exist.

Yet it is not, of course, so simple. All right, if you put the *straight* question, the straight answer Nature gives is 'No'. But the fact is that most people (either in or outside the fields of parapsychology and religion) usually do not ask straight questions, or, even if they do, they do not insist on getting straight answers. They tend instead to ask, for example: 'Do things *sometimes* happen *consistent with the idea of*, or *which would not rule out the possibility of* supernormal powers of cognition or survival after death?' And the answer Nature tends to give back is not a straight 'No', but a 'Maybe', or a 'Sort of', or—rather like a politician—a 'Well, yes and no'. In fact, sometimes Nature behaves even more like a politician. She, or whoever is acting for her, instead of saying 'No', sidesteps the question and says 'Yes' to something else. 'Can you contact my dead uncle?' 'Well, I don't know about that, but I can tell you what's written in this sealed envelope, or I can make this handkerchief vanish.'

I thought the thing to do, then, must be to analyse some of these less than straightforward interchanges—as they occur both in ordinary life and in the parapsychological laboratory—to see what meanings people put on them.

However, not everyone responds to the same tricky evidence in the same way. So I realized it would be important to ask about personal differences: why some individuals (though it has to be said not many) remain pernickety and sceptical while others jump immediately to the most fantastic of conclusions. Thus, in the event, I turned my research to the study of the *psycho-biography* of certain 'extreme believers'—those who throughout history have made the running as evangelists for what I have called 'paranormal fundamentalism'. Who were—and are—these activists, and what's got into them?

Here I'll review just one case history, to see what can be learned. It is, for sure, a special case, but one which touches on many of the wider issues. The case is that of Jesus Christ.[4]

I have several reasons for choosing to discuss the case of Jesus. First, of course, he needs no introduction. Even those who know next to nothing about other heroes of the supernatural, know at least something about Jesus. Second, I think it can be said that the miracles of Jesus, as recorded in the Bible, have done more than anything else to set the stage for all subsequent paranormal phenomena in Western culture, outside as well as inside a specifically religious context. Modern philosophy is not quite, as Whitehead once remarked, merely footnotes to Plato, and modern parapsychology is not quite footnotes to the Bible. But there can be no question that almost all of the major themes of parapsychology do in fact stem from the Biblical tradition. Third, and most important, I think Jesus is probably the best example there has ever been of a person who not only believed in the reality of paranormal powers, but believed *he himself had them.* Jesus, I shall argue, quite probably believed he was the real thing: believed he really was the Son of God, and that he really was capable of performing supernatural miracles. I am going to ask, why?

Now, no doubt one perfectly adequate answer to this question—and a good many people's preferred answer—would be that Jesus believed he was the real thing because *he really was the real thing.* Except that there are, I would say, very strong reasons for supposing that *he really was not the real thing.* And I do not mean merely to point to the reasons that are usually trotted out by sceptics: namely, that it would have been scientifically impossible for him to have been the real thing, totally unprecedented, and so on. I mean to point to the softer but just as devastating grounds that are to be found in the Bible itself. For, when it comes down to it, Jesus just does not seem to have behaved *enough like the real thing.* His supernatural powers (even as recorded by friends) simply were not at the level we should expect of them: they were in fact surprisingly restricted, and not only restricted but restricted in a very suggestive way. Not to put too fine a point on it, Jesus in most if not all of his public demonstrations behaved *just like a conjuror.*

From the point of view of Bible scholarship, I am not saying anything dramatically new here; but, as you'll see, I shall be giving what has already been said a rather different emphasis.

Many Biblical scholars have noted (to their dismay or glee, depending on which side they were taking) that Jesus's miracles were in fact entirely typical of the tradition of performance magic that flourished throughout the ancient world.[5]

Lucian, a Roman born in Syria, writing in the second century, catalogued the range of phenomena that, as he put it, 'the charlatans and tricksters' could lay on, and that were, he said, on display in every market place around the Mediterranean. They included walking on water, materialization and de-materialization, clairvoyance, expulsion of demons, and prophecy. And he went on to explain how many of these feats were achieved by normal means. He described, for example, a 'pseudo-miracle worker' called Marcus who regularly turned water into wine by mixing the water in one cup with red liquid from another cup while the onlookers' attention was distracted.[6]

Christian apologists were, early on, only too well aware of how their Messiah's demonstrations must have looked to outsiders. They tried to play down the alarming parallels. There is even some reason to think that the Gospels themselves were subjected to editing and censorship so as to exclude some of Jesus' more obvious feats of conjuration.[7]

The Christian commentators were, however, in something of a dilemma. They obviously could not afford to exclude the miracles from the story altogether. The somewhat lame solution, adopted by Origen and others, was to admit that the miracles would indeed have been fraudulent if done by anybody else simply to make money, but not when done by Jesus to inspire religious awe. Origen wrote:

The things told of Jesus would be similar to those of the magicians if Celsus had shown that Jesus did them as the magicians do, merely for the sake of showing off. But as things are, none of the magicians, by the things he does, calls the spectators to moral reformation, or teaches the fear of God to those astounded by the show.[8]

Yet there is a question that is not being asked here. Why ever should Jesus have put his followers in this position of having to defend him against these accusations in the first place? If Jesus, as the Son of God, really did have the powers over mind and matter that he claimed, it should surely have been easy for him to have put on entirely different class of show.

Think about it. If a fairy godmother gave *you* this kind of power, what would you do with it? No doubt, you would hardly know where to begin. But, given all the wondrous things you might contrive, would you consider for a moment using these powers to mimic ordinary conjurors: to lay on magical effects of the kind that other people could lay on *without* having them? Would you produce rabbits from hats, or make handkerchiefs disappear, or even saw ladies in half? Would you turn tables, or read the contents of sealed envelopes, or contact a Red Indian guide? No, I imagine you would actually take pains to distance yourself from the common conjurors and small-time spirit mediums, precisely so as not to lay yourself open to being found guilty by association.

I am not suggesting that all of Jesus's miracles were quite of this music-hall variety (although wine into water, or the finding of a coin in the mouth of a fish, are both straight out of the professional conjuror's canon). But what has to be considered surprising is that any of them were so; moreover, that so few of them, at least of those for which the reports are even moderately trustworthy, were altogether of a different order.

With all that power, why can you do *this*, but not do *that*? It seems to have been a common question put to Jesus even in his lifetime. If you are *not* a conjuror, why do you behave so much like one? Why, if you are so omnipotent in general, are you apparently so impotent in particular? Why—and this seems to have been a constant refrain and implied criticism— can you perform your wonders *there* but not *here*?

One of the telltale signs of an ordinary magician would be that his success would often depend on his being able to take advantage of surprise and unfamiliarity. And so, when the people of Jesus's home town, Nazareth, asked that 'whatsoever we have heard done in Capernaum, do also here in thy country',[9] and when Jesus failed to deliver, they were filled with wrath and suspicion and told him to get out. 'And he could there do no mighty work,' wrote Mark.[10] Note 'could not', not 'would not'. Textual analysis has shown that it was a later hand that added to Mark's bald and revealing statement the apologetic rider, 'save that he laid his hands upon a few sick folk, and healed them'.[11]

The excuse given on this occasion in Nazareth was that Jesus's powers failed 'because of their unbelief'.[12] But this, if you think about it, was an oddly circular excuse. Jesus himself acknowledged, even if somewhat grudgingly, that the most effective way to get people to believe in him was to show them his miraculous powers. 'Except ye see signs and wonders,' he admonished his followers, 'ye will not believe.'[13] How then could he blame the fact that he could not produce the miracles they craved on the fact that they did not believe to start with?

Did Jesus himself know the answer to these nagging doubts about his paranormal powers? Did he know why, while he could do so much, he was still unable to do all that other people—and maybe he himself—expected of him?

Remember the taunts of the crowd at the crucifixion: 'If thou be the Son of God, come down from the cross . . . He saved others; himself he cannot save.'[14] Hostile as these taunts were, still they must have seemed even to Jesus like reasonable challenges. We do not know how Jesus answered them. But the final words from the cross, 'My God, my God, why hast thou forsaken me?', suggest genuine bewilderment about why he could not summon up supernatural help when he most needed it.

Why this bewilderment? Let's suppose, for the sake of argument, that Jesus was in fact regularly using deception and trickery in his public performances. Let's suppose that he really had no more paranormal powers than anybody else, and this meant in effect he had no paranormal powers at all. Why might he have been deluded into thinking he was able *genuinely* to exert the powers he claimed?

We should begin, I think, by asking what there may have been in Jesus's personal history that could provide a lead to what came later. It seems pretty clear that Jesus's formative years were, to say the least, highly unusual. Everything we are told about his upbringing suggests that even in the cradle he was regarded as a being apart: someone who, whether or not he was born to greatness, had greatness thrust upon him from an early age.

Admittedly, it is the privilege of many a human infant to be, for a time at least, the apple of his or her parents' eyes. So that the fantasy of being a uniquely favoured human being is actually quite common among little children. For a good many it is a fantasy that is based squarely in the reality of their family relationships. The psychoanalyst Ernst Becker wrote:

> The small child lives in a situation of utter dependence; and when his needs are met it must seem to him that he has magical powers, real omnipotence. If he experiences pain, hunger or discomfort, all he has to do is to scream and he is relieved and lulled by gentle, loving sounds. He is a magician and a telepath who has only to mumble and to imagine and the world turns to his desires.[15]

Although most children must, of course, soon discover that their powers are not really all that they imagined, for many the idea will linger. It seems to be quite usual for young people to continue to speculate about their having abilities that no one else possesses. And it is certainly a common dream of adolescents that they have been personally cut out to save the world.

The fact that the young Jesus may have had intimations of his own greatness might not, therefore, have made him so different from any other child. Except that there were in his case other—quite extraordinary—factors at work to feed his fantasy and give him an even more exaggerated sense of his uniqueness and importance. To start with, there were the very special circumstances of his birth.

Among the Jews living under Roman rule in Palestine at the beginning of the Christian era, it had been long been prophesied that a Messiah, descended from King David, would come to deliver God's chosen people from oppression. And the markers—the tests—by which this saviour should be recognized were known to everybody. They would include: (i) he would indeed be a direct descendant of the king: 'made of the seed of David'.[16] (ii) He would be born to a virgin (or, in literal translation of the Hebrew, to a young unmarried woman): 'Behold, a virgin shall conceive, and bear a son, and shall call his name Immanuel.'[17] (iii) He would emanate from Bethlehem: 'But thou, Bethlehem, though thou be little among the thousands of Judah, yet out of thee shall he come forth unto me that is to be ruler in Israel.'[18] (iv) The birth would be marked by celestial sign: 'A star shall come forth out of Jacob, and a sceptre shall rise out of Israel.'[19]

We cannot of course be sure how close the advent of Jesus actually came to meeting these criteria. The historical facts have been disputed, and many modern scholars would insist that the story of the nativity as told in Matthew's and Luke's Gospels was largely a post hoc reconstruction.[20] Nonetheless, there is, to put it at its weakest, a reasonable possibility that the gist of the story is historically accurate. (i) Even though the detailed genealogies cannot be trusted, it is quite probable that Joseph was, as claimed, descended from David. (ii) Even though it is highly unlikely that Mary was actually a virgin, she could certainly have been carrying a baby before she was married; and, if the father was not Joseph, Mary might—as other women in her situation have been known to—have claimed that she fell pregnant spontaneously. (iii) Granted that Jesus's family were settled in Nazareth, there are still several plausible scenarios that would place his actual birth in Bethlehem (even if Luke's story about the tax census does not add up). (iv) Although the exact date of Jesus's birth is not known, it is known that Halley's comet appeared in the year 12 BC; and, given that other facts suggest that Jesus was born between 10 and 14 BC, there could have been a suitable 'star'.

Some sceptics have felt obliged to challenge the accuracy of this version of events on the grounds that the Old Testament prophets could not possibly have 'foretold' what would happen many centuries ahead. But such scepticism is, I think, off target. For the point to note is that there need be nothing in itself miraculous about foretelling the future, provided the prophet has left it open as to when and where the prophecy is going to be fulfilled. Given that a tribe of, let's say, a hundred thousand people could be expected to have over a million births in the course of, let's say, three hundred years, the chances that one of these births might 'come to pass' in more or less the way foretold are relatively high. It is not that it would *have* to happen to someone somewhere, it is just that, if and when it did happen to someone somewhere, there would be no reason to be too impressed.

The further point to note, however, is that even though there may be nothing surprising about the fact that lightning, for example, strikes somebody somewhere, it may still be very surprising to the person whom it strikes. While nobody is overly impressed that someone or other wins the lottery jackpot every few weeks, there is almost certainly some particular person who cannot believe his or her luck—somebody who cannot but ask: 'Why me?' For the winner of the lottery herself and her close friends, the turn of the wheel will very likely have provided irresistible evidence that fate is smiling on her.

So too, maybe, with the family of Jesus. Suppose that through a chapter of accidents—we need put it no more strongly than that—the birth of Jesus to Joseph and Mary really did meet the preordained criteria. Assume that the set of coincidences was noticed by everyone around, perhaps harped on especially by Mary for her own good reasons, and later drawn to the young boy's attention. Add to this an additional stroke of fortune (which, as told in Matthew's gospel, may or may not be historically accurate): namely, that Jesus escaped the massacre of children that was ordered by Herod, so that he had good reason to think of himself as a 'survivor'. It would seem to be almost inevitable that his family—and later he himself—would have read deep meaning into it, and that they would all have felt there was a great destiny beckoning.

The image a person has of himself is bound to have a crucial influence on his psychological development. Not only will it shape his choices as to what he attempts to make of his own life, but—because of that—it will frequently be self-confirming. The man who believes himself born to be king will attempt to act like a king. The man who knows himself to have been selected from all the possible Tibetan babies to be the future Dalai Lama will allow himself to grow into the part. Children of Hollywood parents who are pushed towards fame and fortune on the stage will sometimes be dramatically successful (provided they are not, by having too much of it, dramatically hurt).

This kind of moulding of a person's character to match an imposed standard can be effective even when the pressures are relatively weak. A surprising, but possibly apposite, example is provided by a recent finding in the area of astrology. When the psychologist Hans Eysenck looked to see whether there is any correlation between particular individuals' 'sun signs' and their personality, he discovered that people born under the odd numbered signs of the zodiac (Aries, Gemini, Leo, Libra, Sagittarius, Aquarius) do in fact tend to be more extrovert than those born under the even-numbered signs (Taurus, Cancer, Virgo, Scorpio, Capricorn, Pisces)—just as predicted by astrologers.[21] But the explanation for this curious finding is almost certainly not that people are being directly influenced by astral forces. A much more likely explanation is that enough people regularly read their horoscopes for the astrologers' predictions to have had a profound effect on what individuals *expect* about their own character and the ways they behave. Further studies by Eysenck showed that the correlation is, in fact, absent in those (relatively few) adults who profess to know nothing about their sun sign or what it 'ought' to mean for them; and it is also absent in children.[22]

Now, if an individual's psychological development can be influenced merely by reading a newspaper horoscope, imagine how it might affect a young man for him to have been born, as it were, under the 'sign of the Messiah'. It would be wholly predictable that, as the meaning dawned on him, it would turn his head to some degree.

The accounts of Jesus's youth do in fact tell of a boy who, besides being highly precocious, was in several ways some-what full of himself and even supercilious.[23] According to the book of James (a near-contemporary 'apocryphal Gospel', supposedly written by Jesus's brother), Jesus during his 'won-drous childhood' struck fear and respect into his playmates by the tricks he played on them. Luke tells the revealing story of how, when Jesus was twelve, he went missing from his family group in Jerusalem and was later discovered by his worried parents in the Temple, 'sitting in the midst of the doctors, both hearing them, and asking them questions'. His mother said: 'Son, why hast thou thus dealt with us? behold thy father and I have sought thee sorrowing.' To which the boy replied: 'How is it that ye sought me? wist ye not that I must be about my Father's business?'[24]

Still, no one can survive entirely on prophecy and promise. However special Jesus thought himself in theory, it is fair to assume he would also have wanted to try his hand in practice. He would have sought—privately as well as publicly—con-firmation of the reputation that was building up around him. He would have wanted concrete evidence that he did indeed have special powers.

In this Jesus would, again, not have been behaving so dif-ferently from other children. Almost every child probably seeks in small ways to test his fantasies, and conducts minor experiments to see just how far his powers extend. 'If I stare at that woman across the street, will she look round?' (Yes, every so often she will.) 'If I pray for my parents to stop squabbling, will they let up?' (Yes, it usually works.) 'If I carry this pebble wherever I go, will it keep the wolves away?' (Yes, almost always.) But we may guess that Jesus, with his especially high opinion of his own potential, might also have experimented

on a grander scale: 'If I command this cup to break in
pieces . . .', 'If I wish my brother's sandals to fly across the
room . . .', 'If I conjure a biscuit to appear inside the urn . . .'.

Unfortunately, unless Jesus really did have paranormal
powers, such larger-scale experiments would mostly have
been unsuccessful. The facts of life would soon have told him
there was a mismatch between the reality and his ambitions.
Jesus would, however, have had to be a lesser child than obviously he was if he let the facts stand in the way—accepting
defeat and possible humiliation. In such circumstances, the
first step would be to *pretend*. And then, since pretence would
never prove wholly satisfactory, the next step would be to
invent.

Break the cup by attaching a fine thread and pulling . . .
Hide the biscuit in your sleeve and retrieve it from the urn . . .
By doing so, at least you get to see how it would feel *if* you
were to have those powers. But what is more, provided you
keep the way you do it to yourself, you get to see how other
people react *if they believe* you actually do have special powers. And since you yourself have reason to think—and in any
case they are continually telling you—that deep down you
really do possess such powers (even if not in this precise way
on the surface), this is fair enough.

So it might have come about that the young Jesus began to
play deliberate tricks on his family and friends and even on
himself—as indeed many other children, probably less talent-
ed and committed than he was, have been known to do when
their reputation is at stake. Yet, although I say 'even on himself', it hardly seems likely he could successfully have denied
to himself what he was up to. In such a situation you may be
able to fool other people most of the time, but not surely yourself. Pretending is one thing. Deluding yourself is another
thing entirely. It is nonetheless your reputation in your own
eyes that really matters. Jesus's position therefore would not
have been an easy or a happy one.

Imagine what it might feel like: to be pretty sure that you
have paranormal powers, to have other people acclaiming

what you do as indeed evidence that you have these powers, but to know that none of it is for real. The more that other people fall for your inventions, the more surely you would yearn for evidence that you could in fact achieve the same results without having to pretend.

Suppose, then, that one day it were to have come about that Jesus discovered, to his own surprise, that his experiments to test his powers had the desired effect *without* his using any sort of trick at all! What might he have made of it then?

I speak from some experience of these matters (and surely some of you have had similar experiences, too). Not, I should say, the experience of deliberately cheating (at least no more than anyone else), but rather of discovering as a child that certain of my own experimental 'try-ons' were successful when I least expected it.

For example, when I was about six years old I invented the game of there being a magic tree on Hampstead Heath, about half a mile from where we lived in London. Every few days I used to visit the tree, imagining to myself, 'What if the fairies have left sweets there?' Sometimes I would say elaborate spells as I walked over there, although I would have felt a fool if anyone had heard me. Yet, remarkably, not long after I began these visits, my spells began to work. Time after time I found toffees in the hollow of the tree trunk.

Or, for another example, when my brother and I were a bit older we started digging for Roman remains in the front garden of our house on the Great North Road in London. Each night we would picture what we would discover the next day, although we had little faith that anything would come of it. Resuming the dig one morning we did find, under a covering of light earth, two antique coins.

Draw your own conclusions about these particular examples. I have, however, a still better case history to tell, which may arguably provide a closer parallel to Jesus's own story. It concerns a teenage boy, Robert, who had become famous as a 'mini-Geller', a child spoon-bender, and whom I met while taking part in a radio programme about the paranormal. I'll tell it here as I wrote it up for a newspaper some years ago:

He had come with his father to the studio to take part in an experiment on psychic metal-bending. He was seated at a table, on which was a small vice holding a metal rod with a strain-gauge attached to it, and his task was to 'will' the rod to buckle.

Half an hour passed, and nothing happened. Then one of the production team, growing bored, picked up a spoon from a tea-tray and idly bent it in two. A few minutes later the producer noticed the bent spoon. 'Well, I never,' she said. 'Hey, Robert, I think it's worked.' Both Robert and his father beamed with pleasure. 'Yes,' Robert said modestly, 'my powers can work at quite a distance; sometimes they kind of take off on their own.'

Later that afternoon I chatted to the boy. A few years previously Robert had seen Uri Geller doing a television show. Robert himself had always been good at conjuring, and—just for fun—he had decided to show off to his family by bending a spoon by sleight of hand. To his surprise, his father had become quite excited and had told everybody they had a psychic genius in the family.

Robert himself, at that stage, had no reason to believe he had any sort of psychic power. But, still, he liked the idea and played along with it. Friends and neighbours were brought in to watch him; his Dad was proud of him. After a few weeks, however, he himself grew tired of the game and wanted to give up. But he did not want people to know that he had been tricking them: so, to save face, he simply told his father that the powers he had were waning.

Next day something remarkable happened. At breakfast, his mother opened the dresser and found that *all* the cutlery had bent. Robert protested it had nothing to do with him. But his Dad said: 'How did he know—perhaps he had done it unconsciously . . .?' There was no denying the cutlery had bent. And how *does* a person know if he has done something *unconsciously*? It was then that Robert realized he must be genuinely psychic.

Since that time, he had received plenty of confirmation. For example, he had only to think about a clock in the next room stopping, and when his father went in he would find that it had stopped. Or he would lie in bed thinking about the door-bell, and it would suddenly start ringing. He had to admit, however, that he was not exactly in control. And the trouble was that people kept on asking him to do more than was psychically within his power. So just occasionally he would go back to using some kind of trick . . . He would not have called it cheating, more a kind of psychic 'filling-in'. Such was the boy's story. Then I talked to his father. Yes, the boy had powers all right: he had proved it time and again. But he was only a kid, and kids easily lose heart; they need encouraging. So, just occasionally, he—the father—would try to restore his son's confidence by arranging for mysterious happenings around the house . . . He would not have called it cheating either, more a kind of psychic 'leg-up'.

This *folie à deux* had persisted for four years. Both father and son were into it up to their necks, and neither could possibly let on to the other his own part in the deception. Not that either felt bad about the 'help' he was providing. After all, each of them had independent evidence that the underlying phenomenon was genuine.[25]

The purpose of my telling this story is not to suggest any exact parallel with Jesus, let alone to point the finger specifically at anyone in Jesus's entourage, but rather to illustrate how easily an honest person *could* get trapped in such a circle: how a combination of his own and others' well-meaning trickery *could* establish a lifetime pattern of fraud laced with genuine belief in powers he did not have.

Still, I cannot deny that, once the idea has been planted that Jesus became caught up this way, it is hard to resist asking who might conceivably have played the supporting role. His mother? Or, if it continued into later years, John the Baptist? Or, later still, Judas, or one of the other disciples? Or all of them by turns? They all had a great interest in spurring Jesus on.

I do not want to suggest any exact parallel with anyone else either. But there are I think several clues that point to the possibility that just this kind of escalating folly might have played a part in the self-development of several other recent heros of the paranormal.

Many modern psychics—notably the Victorian spiritualist D. D. Home, and in our own times Uri Geller—have been childhood prodigies. In Geller's case, for example, it is reported:

At about the age of five he had the ability to predict the outcome of card games played by his mother. Also he noted that spoons and forks would bend while he was eating . . . At school his exam papers seemed identical to those of classmates sitting nearby . . . Classmates also reported that, while sitting near Geller, watches would move forwards or backwards an hour.[26]

And the same precocity is evident with several of those who have gone on to be, if not psychics themselves, powerful spokesmen for the paranormal. Arthur Koestler, for example, has told of how as a boy he 'was in demand for table-turning sessions in Budapest'.[27]

It is also significant that when, as has frequently happened, celebrated psychics have been discovered to be cheating by non-paranormal means, they have nonetheless maintained that they did not *always* cheat. And it is significant too that their supporters have found this mixture of fake and real powers plausible and even reasonable. Of Geller, Koestler said: 'Uri is certainly 25 percent fraud and 25 percent showman, but 50 percent is real.'[28]

But on top of this there is the remarkable degree of self-assurance that these people have typically displayed. Again and again they have left others with the strong impression that they really do believe in their own gifts. Geller has impressed almost all who have met him—I include myself among them—as a man with absolute faith in his own powers. He comes over as being, as it were, the chief of his own disciples. And it is hard to escape the conclusion that in the past and maybe still today he has had incontrovertible evidence, as he sees it, that he is genuinely psychic.

Let me give a small example: when Geller visited my house and offered to bend a spoon, I reached into the kitchen drawer and picked out a spoon that I had put there for the purpose—a spoon from the Israeli airline with EL AL stamped on it. Geller at once claimed credit for this coincidence: 'You pick a spoon at random, and it is an Israeli spoon! My country! These things happen all the time with me. I don't know how I influenced you.' I sensed, as I had done with Robert, that

Geller was really not at all surprised to find that he had, as he thought, influenced me by ESP. It was as if he took it to be just one more of the strange things that typically happen in his presence and which he himself could not explain *except* as evidence of his paranormal powers—'those oddly clownish forces', as he called them on another occasion.[29]

The suspicion grows that someone else has over the years been 'helping' Geller without his knowing it, by unilaterally arranging for a series of apparently genuine minor miracles to occur in his vicinity. A possible candidate for this supportive role might be his shadowy companion, Shipi Shtrang. Shtrang, Geller's best friend from childhood, later his agent and producer, and whose sister he married, would have had every reason to encourage Geller in whatever ways he could; and Geller himself is on record as saying that his powers improve when Shtrang is around.[30] (Noting the family relationship between Shtrang and Geller, we might note too that John the Baptist and Jesus were related: their mothers, according to Luke, were first cousins.)

Now, maybe this all seems too much and too Machiavellian. If you do not see how these examples have any relevance to that of Jesus, you do not see it, and I will not insist by spelling it out further.

But now I have a different and more positive suggestion to make about why a man as remarkable as Jesus (and maybe Geller too) might have become convinced he could work wonders—and all the more strongly as his career progressed. It is that, in some contexts, the very fact of being remarkable may be enough to achieve semi-miraculous results.

I have as yet said little of Jesus as a 'healer'. My reason for not attending to this aspect of his art so far is that so-called 'faith healing' is no longer regarded by most doctors, theologians, or parapsychologists as being strictly paranormal. Although, until quite recently, the miracle cures that Jesus is reported to have effected were thought to be beyond the scope of natural explanation, they no longer are so.

It is now widely recognized by the medical community that the course of many kinds of illness, of the body as well as of the mind, can be influenced by the patient's hopes and expectations and thus by the suggestions given him by an authority figure whom he trusts.[31] Not only can the patient's own mental state, guided by another, profoundly affect the way he himself perceives his symptoms, but it can also help mobilize his body's immunological defences to help achieve more lasting recovery. It is unfortunately true that only certain sorts of illness benefit, and that in any case the effects are not always permanent—so that the pain or the stiffness or the depression tends to return. But, at least in the short term, the cure can be dramatic.

To say that these cures have a normal explanation, however, is not to deny that they may often rely on the *idea* of a paranormal explanation. In fact, it is often quite clear that they do rely on it. It is for most people, including the healer, extremely hard to imagine how the voice or the touch of another person could possibly bring about a cure unless this other person were to have paranormal powers. It follows that the more the patient believes in these powers, the more he will be inclined to take the suggestions seriously—and the better they will work. Equally, the more the healer himself believes in his powers, the more he can make his suggestions sound convincing—and, again, the better they will work.

The consequence is that a kind of virtuous circle can get established. Success in bringing about a cure feeds back to the healer, boosting both his image in the eyes of the world and his image of himself. And thereafter nothing succeeds like more success. The process must of course be launched in some way. There has to be some degree of faith present initially, otherwise the process cannot be expected to get going. But all that this requires is that there should already be some small reason—however unsubstantiated—why people should consider the healer a special person.

In Jesus's case we can assume that, for some or all of the reasons given earlier, his reputation as a potential miracle worker would in fact have been established early in his career and have run ahead of him wherever he went. He would very likely have found, therefore, that he had surprisingly well-developed capacities for healing almost as soon as he first attempted it. And thereafter, as word spread, he would have got better at it still.

Thus, even if I am right in suggesting that his own or others' subterfuges played some part in creating the general mystique with which he was surrounded, we may guess that in this area Jesus would soon have found himself being given all the proof he could have asked for that he was capable of the real thing.

A minor but instructive parallel can again be provided by the case of Uri Geller. As far as I know, Geller has never claimed to be able to heal sick human beings, but he has certainly claimed to be able to heal broken watches. In fact, this was one of the phenomena by which he originally made his name.

In a typical demonstration, Geller would ask someone to provide him with a watch that had stopped working and then, merely by grasping it in his hands for a minute or so, he would set it going again. In his own words (as ever, disarmingly ingenuous): 'I put an energy into it. I don't know what kind of energy it is, but apparently it fixes the watch. I don't think it will ever stop again. Maybe it will, but it's working because I'm around now.'[32]

As with human healing, this kind of watch-cure may in fact be perfectly genuine. It works when it does because the commonest cause of a watch breaking down is that it has become jammed with dust and thickened oil; and, if such a watch is held in the hand for a few minutes, body heat can warm and thin the oil and free the mechanism. When the psychologists David Marks and Richard Kammann, investigating the phenomenon, collected an unbiased sample of sixty-one broken watches and subjected them merely to holding and handling, they found that 57 per cent were successfully started.[33] As with human healing, the cure does not work if there are actually parts broken or missing, nor is it usually permanent. But the effect is unexpected and impressive, all the same.

The cure can easily *seem* paranormal. So, it is not surprising that few if any people watching Geller realized they could in fact perfectly well do it on their own without Geller's encouragement. What probably occurred, therefore, was another example of positive feedback. The higher were people's expectations of Geller's powers—based, as with Jesus, on his preceding reputation as a psychic—the more likely they were to rely on him to give the lead with his suggestions, and hence the more successes he had and the more his reputation spread.

But the crucial question now is what Geller himself thought of it. Is it possible that he was as impressed by his own achievements as everybody else? Suppose he himself had no more understanding of how watches work (at least until it was forcibly brought to his notice by sceptics) than Jesus had of how the immune system works. In that case, he could easily have deduced that he really did have some kind of remarkable power. Like Jesus, he would have been finding that in one area he could genuinely meet his own and others' expectations of him.

And yet . . . And yet, 'about the ninth hour Jesus cried with a loud voice, saying, My God, my God, why hast thou forsaken me?'. Many have speculated on what he meant. But if at the point of death Jesus really did speak these opening lines of the Twenty-Second Psalm, the clue may lie, I suggest, in the way the psalm continues: 'Why art thou so far from helping me? . . . Thou art he that took me out of the womb; thou didst make me hope when I was upon my mother's breasts.'[34]

It was Jesus's tragedy—his glory, too—to have been cast in a role that no one could live up to. He did not choose to be taken from the womb of a particular woman, at a particular place, under a particular star. It was not his fault that he was given quite unrealistic hopes upon his mother's breasts. He did his best to be someone nobody can be. He tried to fulfil the impossible mission. He played the game in the only way it

Fig. 26. Hans Holbein, The Body of the Dead Christ in the Tomb (1521). Öffentliche Kunstammlung, Basle.

could be played. And the game unexpectedly played back—
and overtook him.

Dostoevsky, during his travels in Europe, saw in a Basel
church Holbein's brutally natural painting of the *Dead Christ*,
showing the body of the Saviour reduced to a gangrenous slab
of meat upon a table. In Dostoevsky's novel *The Idiot,* Prince
Myshkin confronts a reproduction of the picture in a friend's
house: 'That picture . . . that picture!' Myshkin cries. 'Why,
some people might lose their faith by looking at that picture.'
Then, later in the novel, Myshkin's alter ego, Ippolit, continues:

As one looks at the body of this tortured man, one cannot help asking oneself the peculiar and interesting question: if the laws of nature
are so powerful, then how *can* they be overcome? How can they be
overcome when even He did not conquer them, He who overcame
nature during His lifetime and whom nature obeyed? . . Looking at
that picture, you get the impression of nature as some enormous,
implacable and dumb beast, or . . . as some huge engine of the latest
design, which has senselessly seized, cut to pieces and swallowed
up— impassively and unfeelingly— a great and priceless Being, a
Being worth the whole of nature and all its laws . . . If, on the eve of
the crucifixion, the Master could have seen what He would look like
when taken from the cross, would he have mounted the cross and
died as he did?[35]

Nature as some enormous, implacable and dumb beast . . .
a huge and senseless engine. Here is what has been at issue all
along. *Nature is the enemy.* What Christ's life seemed to
promise, what the paranormal has always seemed to promise
people, is that we, mere human beings, can vanquish this
dumb beast of natural law. What the emptiness of this promise
eventually forces us to recognize is that, when we do treat
Nature as the enemy and seek to conquer her, Nature
inevitably has the last depressing word: 'No.' We put the question, 'Can a man be a god?', and Nature says—as only she
knows how—'No, a man can be only a mortal man'.

Yet perhaps where we go wrong is with our idea of what we would *like* Nature to say. The assumption that we would be better off if Nature did say 'Yes—I surrender, you can have your miracles'.

This is not the point at which to open up a whole new line of argument. But, so as to lighten this discussion at the end, I shall leave you to consider this quite different possibility: that Nature—Mother Nature—is not the enemy at all, but in reality the best ally we could possibly have had. The possibility that the world we know and everything we value in it has come into being *because of* and not *despite* the fact that the laws of nature are as they are. And what is more, the possibility that it may have been an essential condition of the evolution of life on earth that these laws *preclude* the miracles that so many people imagine would be so desirable: action at a distance, thought transfer, survival of the mind after death, and so on. The possibility that the dumb beast—Nature and her laws—is actually something of a Beauty.

I expect you have heard of the so-called 'anthropic principle': the idea that the laws of physics that we live with in our universe are as they are because if they weren't as they are, this universe wouldn't be the kind of place in which human beings could exist. Adjust the laws ever so little, and it would all go wrong: the universe would expand too fast or too slowly, it would become too hot or too cold, organic molecules would be unstable—and we human beings would not be here to observe and seek the meaning of it.

I believe there is a similar story to be told about the paranormal.[36] That's to say, if the laws of nature were to be adjusted to allow for paranormal phenomena, this too would have a disastrous consequences for human beings, fatally compromising their individuality and creativity and undermining the basis of biological and cultural progress. It would not only be the end of life as we know it, but very possibly enough to prevent life ever having started.

If this is right, it means there is a kind of 'Catch 22' about research in parapsychology, at least if you try to live up to the expectations of the gentlemen who endowed the Perrott–Warrick Fellowship. For it means, perhaps, that you can only be asking those questions of Nature—looking for super-normal powers of cognition or action, or the persistence of the mind after death—in a world where Nature is *bound* to answer: 'No, there's nothing to it.' In any world where Nature *could* say 'Yes' to those questions, we would not be here to ask them.

17

Hello, Aquarius!

Londoners may have seen a poster in the tube stations inviting them to telephone the Zodiac Line to Russell Grant. Always a sucker for having my fortune told, I telephoned two 0's and seven 7's. 'Russell Grant here,' said a voice. 'Thank you for calling. I've got a computer which is going to work out your horoscope, personal only to you. The computer is very sensitive. Please code in your date of birth by pressing the keys on your telephone.' I obliged: 27 . . . 03 . . . 43.

'Hello, Aquarius,' said Russell's voice. 'You were born on . . . January . . . 27th . . . 1943.' Well, not quite; but the poster had promised that Russell would tell me 'a couple of personal things just between you and him', and perhaps this was one of them. I listened, all ears. 'Your winning personality will take the world by storm today.' The sun, it seemed, was shining on me. Money, love, health—all, I was told, were set fair. I would even be a good day for philosophical, religious, and legal projects.

Philosophical, religious, and legal projects? I had as it happened given a talk on the radio the previous evening about the philosophical and legal aspects of the trials of animals which took place in the Middle Ages before religious courts. The coincidence was, to say the least, uncanny.

Shaken somewhat, I tried again. 27 . . . 03 . . . 43. This time the computer decided to take my word for it. 'Hello Aries, you were born on . . . March . . . 27th . . . 1943.' But now, sadly, the forecast had changed. I still had my winning personality. But there was no mention now of philosophy, law, or religion. And my prospects were in general decidedly gloomier. 'Dig for victory,' said Russell. 'Get on your bike.' I tried once more.

'Hello, Pisces,' said the voice (either my phone needs mending, or Russell's computer is peculiarly sensitive to my protean nature). Miraculously, philosophy and religion were back on the agenda, *and* digging for victory.

Third time through, however, I was beginning to become slightly more sensitive myself. Why those telltale pauses after each tidbit of information? Why the just-discernible changes of inflection? I began to realize that the computer was playing a game of heads, bodies, and legs, in which snippets of Russell's wisdom were being recombined (at random? surely not!) to give the impression of originality.

Now I don't say it's not fun. But what it makes it less than harmless fun is the multi-layered deceitfulness of the whole thing. First, the pretence that Mr Grant is there on the end of the line (all right, we all know about answering machines, but the illusion is still powerful); second, the pretence that any genuine astrological analysis has been done by either Mr Grant or the computer; third, the pretence that even if it had been done it would have had any validity at all.

Let the buyer beware, it's said. If people want to waste their money in talking to a robot at 46p a minute (at least £1.00 a call), that's their business. But actually it's not that simple. The little old ladies who even now no doubt are calling in for their daily fix of Russell Grant are—as the promoters of this scheme presumably well know—complicated beings with complicated beliefs. And they cannot be counted on to recognize rubbish when they hear it. Russell Grant has already been given a spurious authority by his appearances on breakfast-time television. Astrology in general is covertly legitimized in a host of ways in newspapers and magazines (even the *Financial Times* has an occasional column giving astrological predictions for the markets). The fact is, our culture provides very little defence against the seductions of irrationality.

But the real problem in persuading people to abandon nonsensical beliefs is precisely the one I came up against on my first call. There can be no doubt that the so-called predictions *sometimes work*. And human beings being what they are, they will often consider one surprising success more than a match for a thousand unsurprising failures. Add to that the human tendency to convert apparent failures to successes by re-interpreting the facts, or to change their own behaviour to bring it into line with what has been predicted, and the persistence of superstition is assured.

Who wins? False prophets, I don't doubt, make quite fat profits. If I had money to invest, I'd invest it straight away in Grant & Co. and their computer. But there must, I'm afraid, be a price to be paid by society when high technology—the fruit of human rationality—is used for such stupid ends. Computers gave us Star Wars; it can be no consolation that they are now giving us Star-Gazing as well.

18

Bugs and Beasts Before the Law

On 5 March 1986, some villagers near Malacca in Malaysia beat to death a dog, which they believed was one of a gang of thieves who transform themselves into animals to carry out their crimes. The story was reported on the front page of the London *Financial Times*. 'When a dog bites a man,' it is said, 'that's not news; but when a man bites a dog, that is news.'

Such stories, however, are apparently not news for very long. Indeed, the most extraordinary examples of people taking retribution against animals seem to have been almost totally forgotten. A few years ago I lighted on a book, first published in 1906, with the surprising title, *The Criminal Prosecution and Capital Punishment of Animals*, by E. P. Evans, author of *Animal Symbolism in Ecclesiastical Architecture*, *Bugs and Beasts Before the Law*, and so on.[1] The frontispiece showed an engraving of a pig, dressed up in a jacket and breeches, being strung up on a gallows in the market square of a town in Normandy in 1386; the pig had been formally tried and convicted of murder by the local court. When I borrowed the book from the Cambridge University Library, I showed this picture of the pig to the librarian. 'Is it a joke?' she asked.

No, it was not a joke.[2] All over Europe, throughout the Middle Ages and right on into the nineteenth century, animals were, as it turns out, tried for human crimes. Dogs, pigs, cows, rats, and even flies and caterpillars were arraigned in court on charges ranging from murder to obscenity. The trials were conducted with full ceremony: evidence was heard on both sides, witnesses were called, and in many cases the accused animal was granted a form of legal aid—a lawyer being

Fig. 27. Frontispiece to E. P. Evans, *The Criminal Prosecution and Capital Punishment of Animals* (London: Heinemann, 1906).

appointed at the taxpayer's expense to conduct the animal's defence.

In 1494, for example, near Clermont in France, a young pig was arrested for having 'strangled and defaced a child in its cradle'. Several witnesses were examined, who testified that 'on the morning of Easter Day, the infant being left alone in its cradle, the said pig entered during the said time the said house and disfigured and ate the face and neck of the said child . . . which in consequence departed this life'. Having weighed up the evidence and found no extenuating circumstances, the judge gave sentence:

We, in detestation and horror of the said crime, and to the end that an example may be made and justice maintained, have said, judged, sentenced, pronounced and appointed that the said porker, now detained as a prisoner and confined in the said abbey, shall be by the master of high works hanged and strangled on a gibbet of wood.[3]

Evans's book details more than two hundred such cases: sparrows being prosecuted for chattering in church, a pig executed for stealing a communion wafer, a cock burnt at the stake for laying an egg. As I read my eyes grew wider and wider. Why did no one tell us *this* at school? We all know how King Canute attempted to stay the tide at Lambeth. But who has heard of the solemn threats made against the tides of locusts which threatened to engulf the countryside of France and Italy?

In the name and by virtue of God, the omnipotent, Father, Son and Holy Spirit, and of Mary, the most blessed Mother of our Lord Jesus Christ, and by the authority of the holy apostles Peter and Paul . . . we admonish by these presents the aforesaid locusts . . . under pain of malediction and anathema to depart from the vineyards and fields of this district within six days from the publication of this sentence and to do no further damage there or elsewhere.[4]

The Pied Piper, who charmed the rats from Hamelin, is a part of legend. But who has heard of Bartholomew Chassenée, a French jurist of the sixteenth century, who made his reputation at the bar as the defence counsel for some rats? The rats had been put on trial in the ecclesiastical court on the charge of having 'feloniously eaten up and wantonly destroyed' the local barley. When the culprits did not in fact turn up in court on the appointed day, Chassenée made use of all his legal cunning to excuse them. They had, he urged in the first place, probably not received the summons, since they moved from village to village; but even if they had received it, they were probably too frightened to obey, since as everyone knew they were in danger of being set on by their mortal enemies, the cats. On this point Chassenée addressed the court at some length, in order to show that if a person be cited to appear at a place to which he cannot come in safety, he may legally refuse. The judge, recognizing the justice of this claim, but being unable to persuade the villagers to keep their cats indoors, was obliged to let the matter drop.

Every case was argued with the utmost ingenuity. Precedents were bandied back and forth, and appeals made to classical and biblical authority. There was no question that God himself—when he created animals—was moving in a most mysterious way, and the court had to rule on what His deeper motives were. In 1478, for example, proceedings were begun near Berne in Switzerland against a species of insect called the inger, which had been damaging the crops. The animals, as was only fair, were first warned by a proclamation from the pulpit:

Thou irrational and imperfect creature, the inger, called imperfect because there was none of thy species in Noah's ark at the time of the great bane and ruin of the deluge, thou art now come in numerous bands and hast done immense damage in the ground and above the ground to the perceptible diminution of food for men and animals; . . . therefore . . . I do command and admonish you, each and all, to depart within the next six days from all places where you have secretly or openly done or might still do damage.[5]

Experience had shown however that the defendants were unlikely to respond:

In case, however, you do not heed this admonition or obey this command, and think you have some reason for not complying with them, I admonish, notify and summon you in virtue of and obedience to the Holy Church, to appear on the sixth day after this execution at precisely one o'clock after midday at Wifflisburg, there to justify yourselves or answer for your conduct through your advocate before his Grace the Bishop of Lausanne or his vicar and deputy. Thereupon my Lord of Lausanne will proceed against you according to the rules of justice with curses and other exorcisms, as is proper in such cases in accordance with legal form and established practice.[6]

The appointed six days having elapsed, the mayor and common council of Berne appointed

after mature deliberation . . . the excellent Thüring Fricker, doctor of the liberal arts and of laws, our now chancellor, to be our legal delegate . . . [to] plead, demur, reply, prove by witnesses, hear judgment, appoint other defenders, and in general and specially do each and every thing which the importance of the cause may demand.[7]

The defender of the insects was to be a certain Jean Perrodet of Freiburg. Perrodet put in the usual plea that since God had created the inger, he must have meant them to survive, indeed to multiply. Was it not stated explicitly in Genesis that on the sixth day of creation God had given 'to every fowl of the air and to everything that creepeth upon the earth . . . the green herbs for meat'?

But the defence in this case was outmatched. The inger, it was claimed in the indictment, were a mistake: *they had not been taken on board Noah's ark.* Hence when God had sent the great flood he must have meant to wipe them out. To have survived at all, the inger must have been illegal stowaways— and as such they clearly had no rights, indeed it was doubtful wheter they were animals at all. The sentence of the court was as follows:

Ye accursed uncleanness of the inger, which shall not be called animals nor mentioned as such . . . your reply through your proctor has been fully heard, and the legal terms have been justly observed by both parties, and a lawful decision pronounced word for word in this wise: We, Benedict of Montferrand, Bishop of Lausanne, etc., having heard the entreaty of the high and mighty lords of Berne against the inger and the ineffectual and rejectable answer of the latter . . . I declare and affirm that you are banned and exorcised, and through the power of Almighty God shall be called accursed and shall daily decrease whithersoever you may go.[8]

The inger did not have a chance. But other ordinary creatures, field-mice or rats, for example, clearly had been present on Noah's ark and they could not be dealt with so summarily. In 1519, the commune of Stelvio in Western Tyrol instituted criminal proceedings against some mice which had been causing severe damage in the fields. But in order that the said mice—being God's creatures and proper animals—might 'be able to show cause for their conduct by pleading their exigencies and distress', a procurator was charged with their defence.[9] Numerous witnesses were called by the prosecution, who testified to the serious injury done by these creatures, which rendered it quite impossible for the tenants to pay their rents. But the counsel for the defence argued to the contrary that the mice actually did good by destroying noxious insects and larvae and by stirring up and enriching the soil. He hoped that, if they did have to be banished, they would at least be treated kindly. He hoped moreover that if any of the creatures were pregnant, they would be given time to be delivered of their young, and only then be made to move away. The judge clearly recognized the reasonableness of the latter request:

Having examined, in the name of all that is just, the case for the prosecution and that for the defence, it is the judgement of this court that the harmful creatures known as field-mice be made to leave the fields and meadows of this community, never to return. Should, however, one or more of the little creatures be pregnant or too young to travel unaided, they shall be granted fourteen day's grace before being made to move.[10]

The trials were by no means merely 'show trials'. Every effort was made to see fair play, and to apply the principles of natural justice. Sometimes the defendants were even awarded compensation. In the fourteenth century, for example, a case

was brought against some flies for causing trouble to the peasants of Mayence. The flies were cited to appear at a specified
time to answer for their conduct; but 'in consideration of their
small size and the fact that they had not yet reached the age of
their majority', the judge appointed an advocate to answer for
them. Their advocate argued so well that the flies, instead of
being punished, were granted a piece of land of their own to
which they were enjoined peaceably to retire.[11] A similar pact
was made with some weevils in a case argued at St-Julien in
1587. In return for the weevils' agreeing to leave the local
vineyards, they were offered a fine estate some distance away.
The weevils' lawyer objected at first that the land offered to his
clients was not good enough, and it was only after a thorough
inspection of it by the court officials that terms were finally
agreed.[12]

In doubtful cases, the courts appear in general to have been
lenient, on the principle of 'innocent until proved guilty
beyond reasonable doubt'. In 1457, a sow was convicted of
murder and sentenced to be 'hanged by the hind feet from a
gallows tree'. Her six piglets, being found stained with blood,
were included in the indictment as accomplices. But no evidence was offered against them, and on account of their
tender age they were acquitted.[13] In 1750, a man and a she-ass
were taken together in an act of buggery. The prosecution
asked for the death sentence for both of them. After due process of law, the man was sentenced, but the animal was let off
on the ground that she was the victim of violence and had not
participated in her master's crime of her own free will. The
local priest gave evidence that he had known the said she-ass
for four years, that she had always shown herself to be virtuous and well behaved, that she had never given occasion of
scandal to anyone, and that therefore he was 'willing to bear
witness that she is in word and deed and in all her habits of life
a most honest creature'.[14]

It would be wrong to assume that, even at the time these
strange trials were going on, everyone took them seriously.
Then as now there were city intellectuals ready to laugh at the

superstitious practices of country folk, and in particular to
mock the pomposity of provincial lawyers. In 1668, Racine
wrote a play—his only comedy—entitled *Les Plaideurs* ('The
Litigants').[15] One scene centres round the trial of a dog in a village of Lower Normandy. The dog has been arrested for stealing a cock. The dog's advocate, L'Intimé, peppers his speeches
with classical references, especially to the *Politics* of Aristotle.
The judge, Dandin, finds it all a bit much. But the defence puts
in a plea of previous good behaviour:[16]

L'INTIMÉ. Fearing nothing, I reach
   The demands of my case and so open my speech.
   Aristotle, in 'Politics' argues it thus,
   And with strength . . .
DANDIN. 'Tis a capon we're here to discuss.
   Aristotle and 'Politics' put on one side.
L'INTIMÉ. But by taking the 'Peripatetic' as guide
   You can prove good and ill . . .
DANDIN. But I rule, Sir, in short,
   Aristotle to be without weight in this court.
   To the point!
L'INTIMÉ. Here's the point. A dog enters a kitchen. That's flat.
   And he finds there a capon deliciously fat.
   He for whom I am pleading with hunger is mad;
   He to whom I'm opposed lies there, plucked, to be had . . .
   And what next, Gentlemen? Why, they come. On what
   score?
   In pursuit of my client. They break in the door,
   And accuse him of theft, nay of brigandage base.
   Then he's dragged by the scruff of the neck in disgrace . . .
   Now if of my client it could be averred
   That he'd eaten the whole or best part of the bird,
   We should claim all the same that his past record still
   Shows sufficient good deeds to make up for his ill.
   Can they show one reproof that he's had beyond doubt?
   Who has guarded the house whene'er they have been out?
   Has he ever forgotten to bark at a thief?

The judge, unfortunately, is unmoved by such considerations, and sentences the dog to the gallows. But the defence counsel has not exhausted his resources. In a last plea for mercy he brings into court a litter of puppies and appeals to the judge's paternal feelings:

L'INTIMÉ. Come! . . . poor desolate strays.
 O! come! for your father may soon be no more.
 With your infantile souls, come plead and implore.

 Yes, gentlemen,—see, Sirs—our misery drear.
 We are orphans: restore then our father so dear.
 Our father begot us; for mercy we pray.
 Our father who now . . .
DANDIN. Go away! Go away!
L'INTIMÉ. Our father, good Sirs . . .
DANDIN. Go away!! My poor ears!
 They have wetted all over the place!
L'INTIMÉ. See our tears!
DANDIN. Och! Already o'erwhelmed with compassion I feel.
 See how telling, when timely, is such an appeal!
 I am puzzled indeed: by the facts quite distressed.
 The crime has been proved: the accused has confessed.
 Yet a sentence on him will not leave us content
 If this quiverful here to the poor-house be sent.

The good judge can stand no more; he is a father himself and has bowels of compassion; and besides, he is a public officer and is chary of putting the state to the expense of bringing up the puppies. When finally a pretty girl turns up in court and puts in her pennyworth on behalf of the accused, the judge generously agrees to let him go.

The interesting thing is that this play was written nearly a hundred years *before* the she-ass, for example, was pardoned on account of her honesty and previous good behaviour by a French provincial court. Laughter, it's been said, is the best detergent for nonsense. Not in France, it seems. But perhaps these trials were *not* altogether nonsense. What was going on?

I shall come to my own interpretation in a while. But first, what help can we get from the professional historians? The answer, so far as I can find it, is: almost none. This is all the more extraordinary because it is not as if the historical evidence has been unavailable. The court records of what were indubitably real trials have existed in the archives for some several hundred years. At the time the trials were occurring they were widely discussed. The existence of this material has been known about by modern scholars. Yet in recent times the whole subject has remained virtually untouched. Two or three papers have appeared in learned journals.[17] Now and again a few of the stories have filtered out.[18] But, for the most part, silence. In 1820, the original church painting of the Falaise pig was whitewashed out of sight; and it is almost as if the same has been done to the historical record—as if the authorities have thought it better that we should be protected from the truth.

I do not know the explanation for this reticence, and can only leave it for the historians to answer for themselves. I dare say, however, that one reason for their embarrassed silence has been the lack, at the level of theory, of anything sensible to say. Take, for example, the long and elegant treatment of the subject by W. W. Hyde (tucked away in the *University of Pennsylvania Law Review*, 1916).[19] Hyde clearly knew enough about the facts: but when it came to *explaining* them, he could do no better than to plump for the remarkable suggestion that 'the savage in his rage at an animal's misdeed obliterates all distinctions between man and beast, and treats the latter in all respects as the equal of the former'.[20]

Recently Esther Cohen has been more constructive.[21] Making a distinction which she says (unfairly, in my view) Evans himself ignores, she points out that it would be wrong to see the trials as being all of a kind. There were in fact two species of trial: on the one hand, those where an individual animal was accused of a specific crime, proceeded against in person (as it were), and condemned like an ordinary criminal, and on the other, those where whole groups of natural pests were accused of being a public nuisance, tried *in absentia*, and dealt with—perforce—by more supernatural means.[22] The former, dating back at least to the Dark Ages, were generally conducted by secular courts under the common law and were relatively down-to-earth affairs. The latter, which did not come into their own until the fifteenth century, were instituted by the Church and as time went on became increasingly suffused with theological hot air: they were, Cohen suggests, closely related to the development of witch trials.[23]

Cohen's discussion of the intellectual background to the trials—particularly her survey of the opinions of contemporary scholars—is informative and interesting. I have no wish to disparage it, and yet I do not find it satisfying. She concentrates entirely upon what the trials may have meant to the authorities sitting in Paris or Rome and ignores their significance for ordinary people. Yet, as she herself notes, learned opinion is one thing, but popular belief quite another; and it seems obvious that there was never any close congruence between the bookish ideas of those contemporary thinkers—from Thomas Aquinas to Leibniz—who mulled over the legal and spiritual dimensions of these trials, and the attitudes of the less-educated people who actually carried out the job of putting animals in the dock. Yet, unless we can get down to the level of 'folk psychology', we have really explained very little. What we need, in short, is an idea of what the common people made of it: for we can be quite sure they made *something* of it, and moreover that whatever that something was *made sense*. The question is, can we make sense of what made sense to them?

Now here two schools of cross-cultural psychology divide. One says, 'other people are other' and we must we wary of ever believing we can understand them by reference to ourselves. 'There is nothing easier', Robert Darnton writes, 'than to slip into the comfortable assumption that Europeans [of the past] thought and felt just as we do today—allowing for the wigs and wooden shoes.'[24] I take his point. But equally, I'd suggest, we must be wary of going to the opposite extreme and of assuming that people of medieval Europe were so different from ourselves that it is not worth trying to apply any of the standards of rationality we have today. True enough, when we hear of people dressing up a pig in human clothes and hanging it as a human criminal, it does seem to indicate that they were in some kind of conceptual muddle. But to suggest, for example, that the grown-up people involved really could not tell the difference between animals and human beings cannot be right. Other people may be other, but they are not necessarily stupid.

So here, in full recognition that this kind of speculation can easily be criticized, let me invite you to put yourself in the place of the participants, and see what kind of explanation might make sense. Let us consider, for example, the trial and subsequent execution of a pig for child-murder—a case that seems to have been all too common, and of which Evans mentions thirty-four recorded instances. Presumably such a trial would have been discussed all over town. What might the mother of the child have said about it, or the farmer to whom the pig belonged, or the law officers who arrested the pig and brought her to the gallows?

No doubt their answers, had we got them, would have varied widely. Yet I imagine there would have been consensus on one point at least: the trial of the pig was not a game. It was undertaken for the good of society, and if properly conducted, it was intended to bring *social benefits* to the community—benefits, that is, to human beings. Certainly the trial had social *costs* for human beings. It not only took up a lot of people's time, it actually cost a good deal of hard cash. The accused animal had to be held in gaol and provided like any other prisoner with the 'king's bread'; expensive lawyers had to be engaged for weeks on end; the hangman, alias 'the master of high works', had to be brought into town, and inevitably he would require new gloves for the occasion. Among the records that have survived there are in fact numerous bills:

To all who shall read these letters: Let it be known that in order to bring to justice a sow that has devoured a little child, the bailiff has incurred the costs, commissions and expenses hereinafter declared:
  Item, expenses of keeping her in gaol, six sous.
  Item, to the master of high works, who came from Paris to Meullant to perform the said execution by command and authority of the said bailiff, fifty-four sous.
  Item, for a carriage to take her to justice, six sous.
  Item, for cords to bind and hale her, two sous eight deniers.
  Item, for a new pair of gloves, two deniers.
  Sum total sixty-nine sous eight deniers. Certified true and correct, and sealed with our seal. Signed: Simon de Baudemont, lieutenant of the bailiff of Mantes and Meullant, March 15th, 1403.[25]

Two centuries later, the cost of prosecuting, hanging, and burning a bitch for an act of gross indecency was set at five hundred pounds—to be recovered in this case from her human accomplice's estate.[26] When people go to this sort of trouble and expense, presumably the community that foots the bill expects to reap some kind of advantage from it.

But what kind of advantage? Actually, three relatively simple answers spring to mind, all worth considering. Let's take them one by one.

First, *the elimination of a social danger.* A pig who has killed once may do so again. Hence, by sentencing the pig to death, the court made life safer for everybody else. This is precisely the reasoning that is still used today in dealing, for example, with a rabid dog or a man-eating tiger. Yet the comparison serves only to show that this can hardly be a sufficient explanation of the medieval trials. It's not just that it would seem unnecessary to go to such lengths to establish the animal's culpability; it's that having found her guilty, the obvious remedy would be to knock her quietly on the head. Far from it: the guilty party was made to suffer the public disgrace of being hung up or burned at the stake in the town square. What is more, she was sometimes subjected to whippings or other tortures before being executed. At Falaise in 1386, the sow that had torn the face and arms of a young child was sentenced to be 'mangled and maimed in the head and forelegs' prior to being hanged. Everything suggests that the intention was not merely to be rid of the animal but to *punish* her for her misdeeds.

Yet why punish an animal? Punishments are commonly given as a means of discouraging any repetition of the crime, either by the original culprit or by others. Capital punishment, of course, is in a special class, since its deterrent effects must be assumed to act only on others (the original culprit being now beyond deterring). Still, if word got around about what happened to the last pig that ate a human child, might not other pigs have been persuaded to think twice?

Again we have a parallel in modern times. A gamekeeper who has shot a magpie, for example, will hang up the dead bird in a prominent position on a fence, in the full expectation that other magpies will see it and take note. But if such was indeed the purpose of executing a pig in the town square, you would think that the officers of justice would—as the gamekeeper does—have gone to some lengths to make sure that the lesson was directed where it mattered, namely, at other potentially delinquent pigs. Not at all: plenty of *people* saw what happened, but so far as I can discover no one ever brought another pig to see the execution. Now, maybe, as some churchmen of the period argued, pigs contain the souls of devils, and devils of course may have their own ways of gathering intelligence. Still, the fact is that if the punishment was meant as a lesson, the obvious target for it would seem to have been the local human populace, not pigs at all.

Then was the purpose, after all, to keep people rather than animals in line? Until comparatively recently, the execution of *human* criminals was done in public with the explicit intention of reminding people what lay in store for future lawbreakers. The effect on impressionable human minds was presumably a powerful one. Then why should not the sight of a *pig* on the scaffold for a human crime have had a like effect? Perhaps it might have proved doubly effective, for such a demonstration that *even pigs* must pay the penalty for law-breaking would surely have given any sensible *person* pause.

As an explanation, this might indeed have something going for it. The extraordinary rigmarole in court, the anthropomorphic language of the lawyers, even the dressing up of the convicted animal *as if it were* a person: these would all fall into place, because only if the proceedings had the semblance of a human trial could the authorities be sure that people would draw the appropriate moral lesson. If the trials look to us now—as to Racine—like pieces of grotesque theatre, that is not surprising. For theatre in a sense they may have been: 'morality plays' designed—perhaps with the full acknowledgement of everyone involved—to demonstrate the power of Church and State to root out crime wherever it occurred.

It would make sense, and yet I do not think this is in fact the answer. What impresses me, reading the transcripts, is that in case after case there was an intellectual and emotional commitment to the process that just does not fit with the whole thing being merely a theatrical charade. There is, moreover, a side that I have not yet revealed, and which provides in some respects the strangest chapter in the story.

I said that we should be wary of dismissing the entire phenomenon of animal trials as the product of an irrational 'medieval mind'. Let me give you now the best reason of all for saying this, which is that similar trials took place centuries earlier in one of the most rational societies the world has ever seen—in ancient Greece. Judicial proceedings against animals were regularly conducted before the Athenian law court known as the Prytaneion. They were alluded to by many classical authors, and Plato himself set out the procedure to be followed:

If a draught animal or any other beast kill a person, unless it be in a combat authorized and instituted by the state, the kinsmen of the slain shall prosecute the said homicide for murder, and the overseers of the public lands, as many as may be commissioned by the said kinsmen, shall adjudicate upon the case and send the offender beyond the boundaries of the country.[27]

But here is the most surprising fact of all. It was not only animals that were prosecuted by the Greeks—*lifeless objects* were brought to court as well: a doorpost for falling on a man and killing him, a sword used by a murderer, a cart which ran over a child. The sentence was again banishment beyond the Athenian boundaries, which for the Greeks was literally 'extermination'. Plato again:

If a lifeless thing shall deprive a person of life, provided it may not be a thunderbolt or other missile hurled by a god, but an object which the said person may have run against or by which he may have been struck and slain . . then the culprit shall be put beyond the boundaries, in the same manner as if it were an animal.[28]

Even in cases where the inanimate object was truly not at fault, no mercy was allowed. Thus the statue erected by the Athenians in honour of the famous athlete Nikon was assailed by his envious rivals and pushed from its pedestal; in falling, it crushed one of its assailants. Although the statue had clearly been provoked—some might even say it was acting in its own defence—it was brought before a tribunal, convicted of manslaughter, and sentenced to be cast into the sea.

Now, whatever we may think about the case of pigs, no one can seriously suppose that the purpose of prosecuting a statue was to prevent it toppling again, or to discourage other statues, or even to provide a moral lesson to Athenians. If such a prosecution had a function, which presumably it did, we must look for it on quite another level. And if that is true here, it is true, I suspect, for the whole panoply of trials we have been considering. Indeed, my hunch is that neither this *nor most of the later trials of animals* had anything to do with what we might call 'preventive justice', with punishment, deterrence, or the discouragement of future crime. I doubt in fact that the *future* conduct of objects, animals, or people was in any way the court's concern: its concern was to establish the meaning to society of the culprit's *past* behaviour.

What the Greeks and medieval Europeans had in common was a deep fear of lawlessness: not so much fear of laws being contravened, as the much worse fear that the world they lived in might not be a lawful place at all. A statue fell on a man out of the blue, a pig killed a baby while its mother was at Mass, swarms of locusts appeared from nowhere and devastated the crops. At first sight such misfortunes must have seemed to have no rhyme or reason to them. To an extent that we today cannot find easy to conceive, these people of the pre-scientific era lived every day at the edge of explanatory darkness. No wonder if, like Einstein in the twentieth century, they were terrified of the real possibility that 'God was playing dice with the universe'.

The same anxiety has indeed continued to pervade more modern minds. Ivan Karamazov, having declared that 'Everything is permitted', concluded that were his thesis to be generally acknowledged, 'every living force on which all life depends would dry up at once'.[29] Alexander Pope claimed that 'order is heaven's first law'.[30] And Yeats drew a grim picture of a lawless world:

> Turning and turning in the widening gyre
> The falcon cannot hear the falconer;
> Things fall apart; the centre cannot hold;
> Mere anarchy is loosed upon the world.[31]

Yet the natural universe, lawful as it may in fact have always been, was never in all respects self-evidently lawful. And people's need to believe that it was so, their faith in determinism, that everything was *not* permitted, that the centre *did* hold, had to be continually confirmed by the success of their attempts at explanation.

So the law courts, on behalf of society, took matters into their own hands. Just as today, when things are un-explained, we expect the institutions of *science* to put the facts on trial, I'd suggest the whole purpose of the legal actions was to establish cognitive control. In other words, the job of the courts was to domesticate chaos, to impose order on a world of accidents—and specifically to make sense of certain seemingly inexplicable events by *redefining them as crimes*.

As the courts recognized, a crime would not be a crime unless the party responsible was aware at the time that he or she was acting in contravention of the law. Thus when a pig, say, was found guilty of a murder, the explicit assumption was that the pig knew very well what she was doing. The pig's action, therefore, was not in any way arbitrary or accidental, and by the same token the child's death became explicable. The child had died as the consequence of an act of calculated wickedness, and however awful that still was, at least it made some kind of sense. Wickedness had a place within the human scheme of things, inexplicable accidents did not. We still see it today, for example, in the reaction people show when a modern court brings in a verdict of 'accidental death' : the relatives of the deceased would often rather believe the death was a deliberate murder than that it was 'bad luck'.[32]

I read recently another report in a London newspaper:

A jilted woman who attempted suicide by leaping from a 12th floor window but landed on and killed a street salesman has been charged with manslaughter. Prosecutors in Taipei, Taiwan said 21-year-old Ho Yu-Mei was responsible for the death of the food salesman because she failed to make sure that there was no one below when she jumped. Ho had argued that she thought the man would have moved away by the time she hit the ground. She also said she had threatened earlier to sue the salesman because 'he interfered' with her freedom to take her own life. If convicted, Ho could be imprisoned for two years.[33]

Who says that the medieval obsession with responsibility has gone away?

But it was with dogs as criminals I began, and with dogs as criminals I'll end. A story in *The Times* some years ago told how a dead dog had been thrown by an unknown hand from the roof of a skyscraper in Johannesburg, had landed on a man, and flattened him—the said man having in consequence departed this life. The headline read—oh, how un-newsworthy!—DOG KILLS MAN. I wonder what Plato, Chassenée, or E. P. Evans would have made of that.[34]

POSTSCRIPT

After this essay was written, I was directed, by Peter Howell, to an ancient account of the trial of a wooden statue (an image of the Virgin Mary) for murder, which took place in North Wales in the tenth century. The case is so remarkable and provides such a clear illustration of the human need to find due cause for an accident, that I give it here in full.[35]

In the sixth year of the reign of Conan (ap Elis ap Anarawd) King of (Gwyneth, or) North Wales (which was about AD 946) there was in the Christian Temple at a place called Harden, in the Kingdom of North Wales, a Rood loft, in which was placed an image of the Virgin Mary, with a very large cross, which was in the hands of the image, called Holy Rood; about this time there happened a very hot and dry summer, so dry, that there was not grass for the cattle; upon which, most of the inhabitants went and prayed to the image, or Holy Rood, that it would cause it to rain, but to no purpose; amongst the rest, the Lady Trawst (whose husband's name was Sytaylht, a Nobleman and Governor of Harden Castle) went to pray to the said Holy Rood, and she praying earnestly and long, the image, or Holy Rood, fell down upon her head and killed her; upon which a great uproar was raised, and it was concluded and resolved upon, to try the said image for the murder of the said Lady Trawst, and a Jury was summoned for this purpose, whose names were as follow, viz.

> Hincot of Hancot, Span of Mancot,
> Leech and Leach, and Cumberbeach,
> Peet and Pate, with Corbin of the Gate,
> Milling and Hughet, with Gill and Pughet

who, upon examination of evidences, declare the said Lady Trawst, to be wilfully murdered by the said Holy Rood, and guilty of the murder, and also guilty in not answering the many petitioners; but whereas the said Holy Rood being very old and done, she was ordered to be hanged—but Span opposed that, saying, they wanted rain, and it would be best to drown her—but was fiercely opposed by Corbin, who answered, as she was Holy Rood, they had no right to kill her, but he advised to lay her on the sands of the river below Harden Castle, from whence they might see what became of her, which was accordingly done; soon after which, the tide of the sea came and carried the said image to some low land (being an island) near the walls of the city called Caer Leon (supposed Chester), where it was found the next day, drowned and dead; upon which the inhabitants of Caer Leon, buried it at the place where found, and erected a monument of stone over it, with this inscription:

> 'The Jews their God did crucify,
> The Hardeners theirs did drown;
> 'Cause with their wants she'd not comply,
> And lies under this cold stone.'

The shaft of the Hawarden Cross, which is supposed to be this very monument, may still be seen, standing in a field known as the Rood-Dee, near the racecourse in Chester.

19

Great Expectations: The Evolutionary Psychology of Faith Healing and the Placebo Effect

> I said that the cure itself is a certain leaf, but in addition to the drug there is a certain charm, which if someone chants when he makes use of it, the medicine altogether restores him to health, but without the charm there is no profit from the leaf.
>
> Plato, *Charmides*, 155–6

I too have a story about leaves and charms. My little daughter, Ada, did not encounter stinging nettles until we returned to England from America when she was nearly four years old. We were in the fields near Cambridge. I pointed out a nettle to her, and warned her not to touch. But, to reassure her, I told her about dock leaves: 'If you get stung,' I said, 'then we'll rub the bad place with a dock leaf and it will very soon be better.'

Ten minutes later Ada had taken her shoes and socks off and had walked into a nettle patch. 'Daddy, daddy, it hurts. Dad, do something.' 'It's all right, we'll find a dock leaf.' I made a show of looking for a dock leaf. But then—in the interests of science—I played a trick.

'Oh dear, I can't see a dock leaf anywhere. But here's a dandelion leaf,' I said, picking a dock leaf. 'I wonder if that will work. I'm afraid it probably won't. Dandelion's aren't the same as dock leaves. They just aren't so magic.'

Ada's foot had come up with a nasty rash. I rubbed it with the dock leaf which Ada thought to be a dandelion. 'Ow,

Daddy, it's no better, it still hurts. It's getting worse.' And the rash certainly looked as bad as ever.

'Let's see if we can't find a proper dock leaf.' And we looked some more. 'Ah, here's just what we need,' I said, picking a dandelion leaf. 'This should work.'

I rubbed Ada's foot again with the dandelion leaf which she now believed to be a dock. 'How's it feel now?' 'Well, a little bit better.' 'But, look, the rash is going away'—as indeed it was. 'It does feel better.' And within a couple of minutes there was nothing left to show.

So, dock-leaf magic clearly works. And yet dock-leaf magic is *placebo* magic. Dock leaves, as such, have no pharmacologically relevant properties (any more than do dandelion leaves). Their power to heal depends on nothing other than the reputation they have acquired over the centuries—a reputation based, so far as I can gather, simply on the grounds than that their old English name, *docce*, sounds like the Latin *doctor*, hence *doctor leaf*, and also that they happen providentially to grow alongside nettles.

But father magic clearly works too. Ada, after all, simply took my word for it that what was needed was a dock leaf. And very likely if I had merely blown her foot a kiss or said a spell it would have worked just as well. Maybe father magic is also a placebo.

We should have a definition. Despite this talk of magic, there's every reason to believe that, when a patient gets better under the influence of a placebo, normal physiological processes of bodily healing are involved. But what's remarkable, and what distinguishes placebos from conventional medical treatments, is that with placebos the process of healing must be entirely self-generated. In fact, with placebos no external help is being provided to the patient's body except by way of ideas being planted in her mind.

Let's say, then, that a placebo is a treatment which, while not being effective through its direct action on the body, works when and because:

The SELF-HEAL Fairy

Fig. 28. Cicely Mary Barker, *The Self-Heal Fairy*, from *Flower Fairies of the Wayside* (Blackie, 1948).

- the patient is *aware* that the treatment is being given;
- the patient has a certain *belief* in the treatment, based, for example, on prior experience or on the treatment's reputation;
- the patient's belief leads her to *expect* that, following this treatment, she is likely to get better;
- the *expectation* influences her capacity for self-cure, so as to hasten the very result that she expects.

How common are placebo effects, so defined? The surprising truth seems to be that they are everywhere. Stories of the kind I've just recounted about Ada are not, of course, to be relied on to make a scientific case. But the scientific evidence has been accumulating, both from experimental studies within mainstream medicine and from the burgeoning research on alternative medicine and faith healing. And it shows beyond doubt that these effects are genuine, powerful and remarkably widespread.[1]

Andrew Weil, one of the best-known advocates of alternative medicine, now argues that 'the art of medicine' in general 'is in the selection of treatments and their presentation to patients in ways that increase their effectiveness through the activation of placebo responses'.[2] And he describes in his book, *Spontaneous Healing*, the range of things that he has found from his own experience can do the trick:

Over the years . . . patients have sung the praises of an astonishing variety of therapies: herbs (familiar and unfamiliar), particular foods and dietary regimens, vitamins and supplements, drugs (prescription, over-the-counter, and illegal), acupuncture, yoga, biofeedback, homeopathy, chiropractic, prayer, massage, psychotherapy, love, marriage, divorce, exercise, sunlight, fasting, and on and on . . . In its totality and range and abundance this material makes one powerful point: *People can get better.*[3]

What's more, as Weil goes on, 'people can get better from all sorts of conditions of disease, even very severe ones of long duration'. Indeed, experimental studies have shown that placebos, as well as being particularly effective for the relief of

pain and inflammation, can,, for example, speed wound healing, boost immune responses to infection, cure angina, prevent asthma, lift depression, and even help fight cancer.
Robert Buckman, a clinical oncologist and professor of
medicine, concludes:

Placebos are extraordinary drugs. They seem to have some effect on
almost every symptom known to mankind, and work in at least a
third of patients and sometimes in up to 60 per cent. They have no
serious side-effects and cannot be given in overdose. In short they
hold the prize for the most adaptable, protean, effective, safe and
cheap drugs in the world's pharmacopoeia.[4]

Likewise, another medical authority, quoted in a recent
review in the *British Medical Journal*, dubs placebos 'the most
effective medication known to science, subjected to more clinical trials than any other medicament yet nearly always doing
better than anticipated. The range of susceptible conditions
appears to be limitless'.[5]

'Limitless' may be an exaggeration. Nonetheless, it's fair to
say that just about wherever placebos *might* work, they *do*. In
other words, wherever a capacity for self-cure exists as a *latent
possibility* in principle, placebos will be found to activate this
capacity in practice. It's true that the effects may not always be
consistent or entirely successful. But they certainly occur with
sufficient regularity and on a sufficient scale to ensure that
they can and do make a highly significant contribution to
human health.

And there's the puzzle: the puzzle that I'll try to address in this
essay from the perspective of evolutionary biology. If placebos
can make such a contribution to human health, then *what are
we waiting for*? Why should it be that we so often need what
amounts to *outside permission* before taking charge of healing
our own bodies?

I can illustrate the paradox with one of Weil's case histories.
He describes the case of a woman with a metastatic cancer in
her abdomen who refused chemotherapy and relied instead on

dieting, exercise, and a regime of 'positive thinking' including 'regular meditation incorporating visualization of tumour shrinkage'—following which, to the physicians' astonishment, the tumour completely disappeared.

Weil asks:

What happened in this woman's abdomen that eliminated widely disseminated cancer and restored her internal organs to good health? Her healing system, probably making use of immune mechanisms, was surely responsible; but *why did it not act before?*[6]

Precisely. Why? Why should her bodily immune system be prepared, apparently, to let her die unless and until her mind decided otherwise?

Weil asks the question as a doctor, and his 'why?' is the why of physiological mechanism: 'What happened?' But I myself, as I said, want to take the perspective of an evolutionist, and my 'why?' is the why of biological function: 'Why are we designed this way?'

There are two reasons for thinking that evolutionary theory may in fact have something important to say here. One reason is that the human capacity to respond to placebos must in the past have had a major impact on people's chances of survival and reproduction (as indeed it does today), which means that it must have been subject to strong pressure from natural selection. The other reason is that this capacity apparently involves dedicated pathways linking the brain and the healing systems, which certainly look is if they have been *designed* to play this very role.[7]

I'd say therefore it is altogether likely that we are dealing with a trait that in one way or another has been shaped up as a *Darwinian adaptation*—an evolved solution to a problem that faced our ancestors.

In which case, the questions are: what was the problem? and what is the solution?

I am not the first to ask these questions. Others have suggested that the key to understanding the placebo response lies in understanding its evolutionary history. George Zajicek wrote in the *Cancer Journal* a few years ago: 'Like any other response in the organism, the placebo effect was selected in Darwinian fashion, and today's organisms are equipped with the best placebo effects.'[8] And Arthur and Elaine Shapiro wrote in a book, *The Placebo Effect*:

Does the ubiquity of the placebo effect throughout history suggest the possibility . . . that positive placebo effects are an inherited adaptive characteristic, conferring evolutionary advantages, and that this allowed more people with the placebo trait to survive than those without it?[9]

But, as these quotations illustrate only too well, the thinking in this area has tended to be question-begging and unrevealing. I hope we can do better.

So, let me tell you the conclusion I myself have come to. And then I shall explain how I have come to it, and where it leads.

My view is this. The human capacity for responding to placebos is in fact not necessarily adaptive in its own right (indeed, it can sometimes even be maladaptive). Instead, this capacity is an emergent property of something else that *is* genuinely adaptive: namely, a specially designed procedure for 'economic resource management' that is, I believe, one of the key features of the 'natural health-care service' which has evolved in ourselves and other animals to help us deal throughout our lives with repeated bouts of sickness, injury, and other threats to our well-being.

Now, if you wonder about this choice of managerial terminology for talking about biological healing systems, I should say that it is quite deliberate (and so is the pun on 'National Health Service'). With the phrase 'natural health-care service' I do intend to evoke, at a biological level, all the economic connotations that are so much a part of modern health-care in society. 'Managed health-care' as it's practised these days is of course not just to do with health—sometimes it isn't even *primarily* to do with health—it is to do with balancing budgets, cutting costs, deciding resource allocation, operating a triage system, and so on. I am suggesting that the same applies in crucial ways to nature's own bodily healing systems.

And the point is that, if that's right, we can take a new theoretical approach. Suppose we adopt the point of view not of the doctors or the nurses in a hospital, but of the hospital administrator whose concern as much as anything is to husband resources, spend less on drugs, free up beds, discharge patients earlier, and so on. Then, if we take this view of the natural health-care service, instead of asking about the adaptiveness of bodily healing as such, we can turn the question round and ask about the adaptiveness of features that *limit* healing or *delay* it. What we'll be doing is a kind of figure-ground reversal, looking at the *gaps between* the healing.

So let's try it. And, in taking this approach, let's go about it logically. That's to say, let's start with the bare facts and then try to deduce what else has to be going on behind the scenes to explain these facts, on the assumption that we are indeed dealing with a health-care system that has been designed to increase people's overall chances of survival. Then, once we know what *has to be* the case, we shall surely be well placed to take a closer look at what actually *is* the case.

I am setting this out somewhat formally, and each proposition should be taken slowly. Here are some basic facts to start with:

1. *Other things being equal, well-being is to be preferred to sickness.*

I assume this is uncontroversial.

2. *People's bodies and minds have a considerable capacity for curing themselves.*

This is what all the evidence of spontaneous recovery shows.

3. *Sometimes this capacity for self-cure is not expressed spontaneously, but can be triggered by the influence of a third party.*

This is the basic placebo phenomenon.

Then, it follows from 1 and 2, presumably, that:

4. *In such cases, self-cure is being* inhibited *until the third-party influence* releases *it.*

And from the assumption that this pattern is non-accidental, that:

5. *When self-cure is inhibited there must be good reason for this* under the existing circumstances; *and when inhibition is lifted there must be good reason for this* under the new circumstances.

Put this together with the starting assumption that people want to be as well as they can be, and we have:

6. *The good reason for inhibiting self-cure must be that the subject is likely to be better off, for the time being,* not being cured.

In which case:

7. *Either there must be benefits to remaining sick, or there must be costs to the process of self-cure.*

Likewise:

8. *The good reason for lifting the inhibition must be that the subject is now likely to be better off* if self-cure goes ahead.

In which case:

9. *Either the benefits of remaining sick must now be less, or the costs of the process of self-cure must now be less.*

I'd say all the above do follow deductively. Given the premises, something like these conclusions *must* be true. In

Pretences

which case, our next step ought to be to turn to the real world and to find out how and in what sense these rather surprising conclusions *could* be true.

The following are the most obvious matters which want unwrapping and substantiating:

Is it indeed sometimes the case that there are benefits to remaining sick and, correspondingly, costs to premature cure?

Is it, as we might guess, more usually the case that there are benefits to getting better and, correspondingly, costs to delayed cure?

In either case, are there really costs associated with the process of cure as such?

Is it possible to predict how these costs and benefits will change according to external circumstances, so that the subject might in fact be able to take control of her own health budget?

Let me take these questions in turn.

Benefits of remaining sick vs. costs of premature cure

It depends how we define sickness. If sickness means a *pathological* condition of the body or mind that is unconditionally harmful, then there cannot of course ever be benefits to remaining sick. However, if sickness is taken more broadly to mean any abnormal condition of body or mind that you, the patient, find distressing and from which you seek relief, then it may be quite another matter.

It has been one of the major contributions of evolutionary theory to medicine to remind us that many of those conditions from which people seek relief are not in fact defects in themselves, but rather self-generated *defences* against another more real defect or threat.[10]

Pain is the most obvious example. Pain is not itself a case of bodily damage or malfunction—it is an adaptive response to it. The main function of your feeling pain is to deter you from incurring further injury, and to encourage you to hole up and rest. Unpleasant as it may be, pain is nonetheless generally *a good thing*—not so much a problem as a part of the solution.

It's a similar story with many other nasty symptoms. For example, fever associated with infection is a way of helping you to fight off the invading bacteria or viruses. Vomiting serves to rid your body of toxins. And the same for certain psychological symptoms too. Phobias serve to limit your exposure to potential dangers. Depression can help bring about a change in your lifestyle. Crying and tears signal your need for love or care. And so on.

Now, just to the extent that these evolved defences are indeed *defences against something worse*, it stands to reason that there will be benefits to keeping them in place and costs to premature cure. If you don't feel pain, you're much more likely to exacerbate an injury; if you have your bout of influenza controlled by aspirin, you may take considerably longer to recover; if you take Prozac to avoid facing social reality you may end up repeating the same mistakes, and so on. The moral is: *sometimes it really is good to keep on feeling bad*.[11]

So, in that case, how about the other side?

Benefits of getting better vs. costs of delayed cure

There *is*, of course, another side. Even when an ailment is one of these evolved defences, this does not necessarily mean there is nothing legitimately to complain of. For the fact is that, while the defences are there to do you good, they may still in themselves be quite a burden—not just because you do not like them, but because they can actually threaten your fitness directly, sometimes severely so.

Take the case of pain again. Yes, it helps protect you. Nevertheless, it is by no means without cost. When pain makes it hard to move your limbs you may become more vulnerable to other dangers, such as predators. When the horribleness of pain takes all your attention, you may no longer be capable of thinking clearly. When pain causes psychological stress, it may make you bad-tempered or incapable or hopeless. It may even take away your will to live. In cancer wards it's said that patients in greatest pain are likely to die soonest and that treating the pain with morphine can actually prolong life.

Or take some of the other defences I listed above. Fever helps fight infection, but it also drains your energy and can have damaging side effects such as febrile convulsions. Vomiting helps get rid of toxins, but it also throws away nourishment. Depression helps disconnect from maladaptive situations, but it also leads to social withdrawal and loss of initiative. Crying helps bring rescue from friends, but it also reveals your situation to potential enemies.

So *now* it stands to reason that there will after all be benefits to getting better and costs to delaying cure. The moral is: *sometimes it really is bad not to return to feeling good as soon as possible.*

And I have been discussing here only examples of self-generated defences. In cases where the sickness in question is a genuine defect or malfunction—a broken leg, say, or a snake bite or a cancerous growth—the balance of advantage must be clearer still. Surely in these cases there could be no advantage at all in withholding cure.

Or couldn't there be?

Costs of the process of cure as such

When the sickness is self-generated, so that cure can be achieved simply by switching off whatever internal process is responsible for generating the symptoms in the first place, then, it's true, the cure comes cheap—and there should indeed be little reason to hold back just cost grounds.

With pain, for example, you may well be able to achieve relief, when and if desirable, simply by sending a barrage of nerve signals down your own spinal cord or by releasing a small amount of endogenous opiate molecules. Similarly with depression, you may be able to lift your mood simply by producing some extra serotonin. The production costs of the neurotransmitters, the endorphins or serotonin, are hardly likely to be a serious limitation.

However, it may be a very different story when the sickness involves genuine pathology and the cure requires extensive repair work or a drawn-out battle against foreign invaders— as with healing a wound or fighting an infection or cancer. For in this case the process of cure may turn out to be far from cheap.

In particular, if and when the cure involves the activity of the immune system, the costs can mount rapidly.[12] For a start, the production of immune agents (antibodies and so on) uses up a surprisingly large amount of metabolic energy; so much so, that early in life when a child's immune system is being built up, it actually takes as much energy as does the brain; and it has been found in animals that when the immune system is artificially stimulated into extra activity, the animals lose weight unless they are given extra food.[13] But besides the calories, the production of immune agents also requires a continual supply of quite rare nutrients in the diet, such as carotenoids.[14] Ideally you should be able to build up reserves when times are good. But, even so, once a major immune response is under way, even the best reserves may get used up—so that every time you go on the attack against today's threat you are compromising your ability to respond to a *new* threat further down the road.

And then, as if this weren't enough, there is the added problem that mounting an immune response can pose quite a danger to your own body, because the immune agents, unless they are controlled, may turn on your own tissues and cause autoimmune disease. This is a particular danger when the body is already under stress, for example as a result of strenuous exercise.[15] It means that even when the resources for overcoming an invader are potentially available, it will not always be possible to deploy them safely to maximum effect.

The overall costs of the immune system are in fact so very great that most people most of the time simply cannot afford to keep their system in tip-top order. At any rate this is true for animals living in the wild. And the ramifications of this go beyond health issues as such, to affect courtship and reproduction. In several animal species, and maybe humans too, differential ability to maintain the costly immune system has become the basis for sexual selection.[16] So when a male or female is looking for a mate they pay close attention to indicators of *immunocompetence*—as shown, for example, by the colours of a bird's feathers, or the symmetry of the body or quality of skin.

All in all, we are beginning to see just how complex the accounting has to be. We have found that with self-generated defences, the sickness itself is designed to bring benefits, but it also has costs, although the process of self-cure is cheap. With sickness arising from outside, the sickness has costs with no benefits, but the process of self-cure can be costly.

Imagine, now, you are one of those hospital managers in charge of this natural health service. It would be the height of irresponsibility if you were to allow all the different departments to operate autonomously, with defences and cures being switched on or off regardless. Instead you should surely be trying to plan ahead, so that you are in a position to decide on your *present priorities* in the light of *future needs*. You would want to ask what weight to attach to this cost or that benefit *in the particular circumstances that you're now in*.

Reckoning how the costs and benefits change according to circumstances

Let's think about some particular examples.

Pain. You've sprained your ankle. Question: is this the defence you really need right now, or on this occasion will it actually do more harm than good? Suppose you are chasing a gazelle and the pain has made you stop—then, fair enough, it's going to save your ankle from further damage even if it means your losing the gazelle. But suppose you yourself are being chased by a lion—then if you stop it will likely be the end of you.

Nausea. You feel sick when you taste strange food. Question: how serious a risk would you be taking in your present state of health if you did not feel sick? Suppose you are a pregnant woman, and it's essential to protect your baby from toxins— then, yes, you really don't want to take any risks with what you eat. But suppose you are starving and this is the only food there is—then you and the baby will suffer still more if you don't take the chance.

Crying. You are upset by a letter telling you you've been betrayed. Question: are the benefits of sending out this signal of distress actually on offer in present company? Suppose you are among friends who can be expected to sympathize with you—then, well and good. But suppose you are among strangers—they may merely find your display embarrassing.

Immune response. You have a head cold. Question: is this the right time to mount a full-scale immune response, or should you be saving your reserves in case of something more serious? Suppose you are safe at home with your family and there's going to be time enough to convalesce—then, sure, you can afford to throw everything you've got against the virus. But suppose you are abroad, facing unknown dangers, under physical stress—then, paradoxically, since you might think you'd want to get better as fast as possible, on balance you might do better to keep your response to the level of a holding operation.

As we see, especially with this last example, the crucial question will very often be: *what is going to happen next*?

The fact is that many of the health-care measures we've

been discussing are *precautionary* measures designed to protect from dangers that lie ahead *in an uncertain future*. Pain is
a way of making sure you give your body rest just in case you
need it. Rationing the use of the immune system is a way of
making sure you have the resources to cope with renewed
attacks just in case they happen. Your healing systems are
basically tending to be cautious, and sometimes overcautious, as if working on the principle of better safe than
sorry.

Now, this principle is clearly a sensible one, if and when you
really cannot tell what is in store for you. It's the same principle that advises you to carry an umbrella when you don't
know whether it's going to rain, or to keep your running pace
down to a trot when you don't know how far you may have to
run.

But suppose you *were* to know.

Continuing the analogy, suppose you could tell for sure that
the weather would stay dry—then you could leave the umbrella behind. Or suppose you had definite information that the
race was just one hundred metres long—then you could sprint
all the way.

And it is the same with health care. Suppose you were to be
able to tell in advance that an injury was non-serious, food
was non-toxic, an infection was not going to lead to compli-
cations, no further threats were in the offing, rescue was just
around the corner, tender loving care about to be supplied—
then, on a variety of levels, you could let down your guard.

It will now be clear, I hope, where this is going. I have been
saying 'you' should ask these questions about costs and benefits, as if it were a matter of each individual acting as his or her
own health-care manager at a rational, conscious level. But of
course what I am really leading up to is the suggestion that
Nature has already asked the same questions up ahead and
supplied a set of adaptive answers—so that humans now have
efficient management strategies built into their constitution.

I would expect this to have occurred on two levels.

To start with, given that there are certain universals in how people fare in different situations, there are presumably *general rules* to be found linking *global* features of the physical and psychological environment to changes in the costs and benefits of health-care—features such as where you live, what the weather is like, the season of the year, what you can see out of the window, how much you feel at home here, and, especially important, what social support you've got.

Generally speaking, any such features that make you feel happy and secure—success, good company, sunshine, regular meals, comforting rituals—are going to be associated with lower benefits to having the symptoms of illness (for example, feeling pain) and lower costs to self-cure (for example, mounting a full-scale immune response). By the same token, any of them that make you feel worried and alone—failure, winter darkness, losing a job, moving house—are going to be associated with higher benefits to continuing to show the symptoms and higher costs to self-cure.

Now, appreciating these cost/benefit changes, and switching health-care strategy accordingly, would bring clear advantages in the competition for biological survival. So it's a fair bet that natural selection will have discovered these general rules and that humans will now be genetically designed to make good use of them—with the appropriate environmentally activated switches being wired into the brain. And indeed there is lots of evidence that this is actually the case. People's health does respond quite predictably to global environmental factors of the sort just listed. Much of the science of 'health psychology' is now concerned with charting just how important these global effects are.[17]

However, let's come to the second level. Knowing the general rules as they apply to everybody is all very well. But it should surely be possible to do still better. If only you were to have a more precise understanding of *local* conditions and the rules as they apply in your own individual case, then in principle you should be able to come up with predictions at the level of *personal expectancies*. These then could be used to fine-tune your health-care management policy to even greater advantage. In which case, natural selection—in an ideal world, anyway—should have discovered this as well.

But, in reality, this might not be straightforward. The problem is likely to be that specific personal expectancies must involve high-level cognitive processing, taking account of learned associations, reasons, and beliefs that are peculiar to each individual. And it would surely be asking too much to suppose that human beings could have been genetically designed so that every one-off prediction generated by an individual brain will have just the right effect directly on the healing systems.

If this were asking too much, however, then presumably the answer would have been for natural selection to go for a bit less. And, in any case, I'd suggest that such a degree of direct one-to-one linkage would actually have been something of an extravagance. For the fact is that most of the benefits of personal prediction ought to be achievable in practice by the much simpler expedient of linking expectancy to healing by way of an *emotional variable*: for which there exists a ready-made candidate, in the form of 'hope' and its antithesis 'despair'.

Thus, what we might expect to have evolved—and it would do the job quite well—would be an arrangement on the following lines. Each individual's beliefs and information create specific expectancies as to how things will turn out for them. These expectancies generate specific hopes—or, as it may be, despairs. And these hopes and despairs, being generic human feelings, act directly on the healing system, in ways shared by all human beings. Hope and despair will have become a crucial feature of the *internal environment* to which human individuals have been designed to respond with appropriate budgetary measures.

I am suggesting natural selection could and should have arranged things this way. It would make sense. Yet it remains to be seen whether it is actually the case. Do we have evidence that hope can and does play this crucial role in health care?

Well, yes, of course we do have evidence. It's the very evidence we started with. Ada and the stinging nettles. The lady with the vanishing cancer. Placebos themselves provide as good evidence as we could ask for. Because what placebo treatments do is precisely to give people reason to hope, albeit that the reason may in fact be specious. No matter, it works! People do change their priorities, let down their guard, throw caution to the winds. That's the placebo effect!

I shall say more about placebos in a moment. But let me first bring in some research on the effects of hope that might have been tailor-made for the purpose of testing these ideas (although in fact it was done quite independently). This is the research of my former colleague at the New School, Shlomo Breznitz.

Breznitz begins a recent paper on 'The Effect of Hope on Pain Tolerance'[18] with a discussion of what has come to be called the 'tour of duty phenomenon'. In the Second World War, American bomber crews who flew nightly missions over Germany suffered terrible casualties, with planes and crews being lost each night. Morale was inevitably low, and low morale brought with it a variety of stress-related illnesses. Psychologists were called in to help. They surmised that a main cause of the problem was uncertainty as to when the ordeal would come to an end. Their advice was to give the crews specific information about when their tour of duty would be over.

Forty missions had been the unofficial average. But now the airmen were told it would be forty missions and no more. The effects were dramatic: once the airmen knew that they only had to hold out *just so long*, they became able to cope as never before.

Breznitz in his own research set out to study this phenomenon experimentally. He undertook a series of studies in which human subjects were made to suffer pain or anxiety or fatigue, and he manipulated their expectations about how long it might last. His question was whether hope for a relatively early end would indeed allow subjects to manage better, while dread that the ordeal might go on and on would do the opposite.

In one study, for example, he required subjects to keep one of their hands in ice-cold water until they could no longer stand the pain and had to remove it. Subjects in one group were *told* that the test would be over in four minutes, while those in another group were not told anything. In fact the test lasted a maximum of four minutes in both cases. The result was that 60 per cent of those who knew when the test would end were able to endure the full four minutes, whereas only 30 per cent of those who were kept in the dark were.

Breznitz does not report whether the subjects who were in the know actually felt less pain than the others (this was a study done for the Israeli army, and no doubt the sponsors were more interested in objective behaviour than subjective feeling). But I strongly suspect that, if the question had been asked, it would have been found that, yes, when the subject had reason to believe that the external cause of the pain was shortly to be removed, it actually *hurt less*.

You'll see, I trust, how nicely this could fit with the health-care management story that I have been proposing. My explanation would be that when it's known that the threat posed by the cause of the pain is soon to be lifted, there's much less need to feel the pain as a precautionary defence. Likewise with the tour of duty phenomenon, my explanation would be that when it's known that safety and rest are coming relatively soon, there's much less need to employ defences such as anxiety and, furthermore, healing systems such as the immune response can be thrown into high gear.

I should stress this is *my* take on things, not Breznitz's. But I am glad to say he himself concludes the review of his research with words that might have been my own: 'it is quite conceivable that people are capable of fine-tuning the distribution of their resources according to the anticipated duration ahead of them.'[19]

So I'll turn to how placebos fit with and extend the story. But first there is one more twist—relating to the question of how hopes get generated.

Breznitz's experimental situations are, of course, somewhat contrived. If people in the real world are fine-tuning their health-care budgets on the basis of their specific hopes, we need to consider how they come to have these hopes. Or, because this is what it will often come down to in the cases that interest us, how they come to have the *beliefs* that underpin these hopes—especially the belief in the power of a particular treatment to bring recovery.

Well, we have just seen one way in which you can come to believe something: someone you suppose to be trustworthy *tells* you so. 'You have my word for it, I promise it will be over in four minutes.' But this is presumably not the only way or indeed the usual way.

In general there are going to be three different ways by which you are likely to acquire a new belief (discussed at greater length in my book, *Soul Searching*).[20] These are: (a) *personal experience*—observing something with your own eyes, (b) *rational argument*—being able to reason your way to the belief by logical argument, and (c) *external authority*—coming under the sway of a respected external figure.

Now, if these are indeed the ways you can come to believe that *a specific treatment is good for you*, we should find each of these factors affecting your hopes for your future well-being—and so, presumably, having significant effects on health management strategies.

Do we? Again, I trust it will have been obvious where the argument is going. But now I want to bring *placebos* front stage—not just as corroborative evidence but as very proof of these ideas about hope and health management.

Suppose we were designing a scientific test of the predicted hope effects. We would want to show that treatments that generate hope in each of the ways just described are capable of calling up the healing process. But, to show it's really *hope* that's doing it, we would want to be able to separate the 'pure hope' component of the treatment from any other component that might be having direct effects on healing. So ideally we would want to use 'hope only' treatments. Which is, of course, precisely what pure placebos are.

Let's see, then. How do these three hope-generating factors play?

Personal experience

Someone believes she is going to recover after a treatment because it has worked for her before. 'This pink pill, labelled "aspirin", cured my headache last time. I'll soon be fine when I take it again.' Or, 'When Nurse Jones looked after me before, I found I was in completely safe hands. I'll be safe enough this time too.'

Past associations of this kind are indeed a fairly reliable basis for hopes about the future. There ought then to be a major *learned* component to placebos. And there is. In fact the learning that recovery from sickness has in the past been associated with particular colours, tastes, labels, faces, and so on is the commonest way by which placebo properties get to be conferred on an otherwise ineffective treatment.

Rational argument

Someone believes in a treatment because she reckons she understands something about the causal basis. 'I know that if a medical treatment is to work it must be *strong medicine*, the kind that tastes nasty, or hurts, or is very expensive. This rhinoceros horn extract has all the right properties.' Or, 'I realize I can't expect to get better unless I myself take positive action and *do something*. That's why I am taking a flight to Lourdes.'

Such arguments do make reasonable sense, and provide rational grounds for hope. So there ought to be a major *rationalized* component to placebos. Again there is. In fact, so-called 'active placebos'—for example, those with unpleasant side-effects or those that involve pseudo-medical procedures such as injection or surgery—are found to work especially well; as, equally, do those that require effort and involvement by the patient. Patrick Wall, the leading pain researcher, has concluded that the process of *doing something about the pain* is indeed all there is to placebo analgesia.[21]

External authority

Someone believes in a treatment because she trusts the word of someone else. 'Everyone I know swears by homeopathy. Presumably it will work for me as well.' 'The doctor has a degree from Harvard. Obviously he'll give me the best treatment going.'

In general, taking the word of others with some claim to authority is a sensible strategy, and people would be foolish not to let their hopes be swayed this way. So there ought to be a major *faith* component to placebos. And once again, of course, there is. In fact, the evidence ranging from faith healing to hypnosis shows that this is potentially the most effective way of all by which otherwise bland treatments and procedures can achieve marvellous powers.

Enough. And it's time to take stock. Let me run back over how the argument has gone so far.

We began by asking what the existence of placebos tells us about the nature of our healing systems. We showed, deductively, that placebos imply that there has to be some kind of health-care management policy in place. And we discovered that, when we look into the biology and natural history, the existence of such a policy would make sense as an evolved adaptation. We speculated about how natural selection would have designed it. We came up with some ideas about the cognitive factors that might well be used to fine-tune the allocation of resources. We predicted what would follow—and asked for evidence. And I think it's fair to say we've found it. The circle is closed. The facts about placebos are evidence for the very analysis that placebos started.

Yet there is some unfinished business here—rather serious and interesting business. And it will become clear just what the issue is as soon as we try, as we probably should have done already, to straighten out a potential confusion about terminology.

The term 'placebo', as we defined it at the outset, was supposed to apply only to those treatments that do nothing *directly* to help with healing, but rather exert their influence on bodily processes *indirectly*, through what we've been calling the 'placebo effect' or 'placebo response'. By contrast, if the treatment does have a direct effect on healing, we should consider it to be a 'genuine treatment' and not a 'placebo'.

However, it hardly needs saying, at this stage of this essay, that genuine treatments with direct effects can of course *also* influence healing indirectly. In fact genuine treatments—for obvious reasons, to do with how easy it is for the subject to believe in them—are even more likely than are placebos to bring about the kind of hope-based changes in health-care strategy that we've been considering. Thus, odd though this may sound, we should want to say that genuine treatments often have placebo effects.

Now, to avoid having to talk this way, I can see we might do better to introduce an entirely new term for these effects: perhaps something like 'hope-for-relief effects', as the generic name for the hope-based components of *any* kind of treatment. Except that this is clumsy and also arguably too theory-ridden. And in any case, the reality is that in the placebo literature the term 'placebo effect' has already become to some degree established in this wider role. So let's keep to it, provided we realize what we are doing.

However, in that case there is surely a further distinction that we should want to draw: namely, between what we may call 'justified' and 'unjustified placebo effects', based on *valid* and *invalid* hopes.

Suppose, for example, a doctor gives someone who is suffering an infection a pill that the patient *rightly* believes to contain an antibiotic: because the patient's hopes will be raised, she will no doubt make appropriate adjustments to her health management strategy—lowering her precautionary defences and turning up her immune system in anticipation of the sickness not lasting long. But now suppose instead he gives her a pill that she *wrongly* believes to contain an antibiotic: because her hopes will be raised in the same way, she will no doubt again make adjustments—in fact, the very same ones. In the first case it would be a justified placebo response, in the second an unjustified placebo response.

I said we should want to draw this distinction, but actually why should we? For it might be argued that, as we have just seen, it will not be a distinction of any consequence. From the point of view of the subject, all hopes are on a par—all hopes have *subjectively* to be valid or else they would not count as hopes. So, provided the patient truly believes that when the doctor gives her the pill she can expect a quick recovery, it is not going to make any difference to her response whether it's the real thing or the fake.

However, to argue this would be to have missed the main point of this essay. For it ought to be clear that, on another level, the distinction may be crucial: the reason being that only when the patient's hope is valid will her anticipatory adjustments to her healing system be likely to be *biologically adaptive*. In fact, when her hope is invalid the same adjustments may actually be *maladaptive*—because she may be risking lowering her fever too soon or using up her precious resources when she cannot afford to.

Thus, from the point of view of natural selection, all hopes are by no means on a par. Unjustified placebo responses, triggered by invalid hopes, must be counted a *biological mistake*.

Why, then, haven't humans evolved *not* to make such mistakes?

The answer, most likely, is that this has never really been an option. The possibility of making mistakes comes with the territory. If you have evolved to be open to true information about the future—coming from experience, reason, or

authority—you are bound to be vulnerable to the disinforma-
tion that sometimes comes by the same routes. You cannot
reap the advantages without running the risks.

Still, the chance of encountering fakes in the natural world
in which human beings evolved was presumably relatively
small—relative, at any rate, to the world we now are in. So
placebos were probably hardly an issue for most of human
prehistory.

This is not to say that in the past the risk did not exist at all.
Superstition has always existed. Indeed it is a pre-human trait.
And when I hear of chimpanzees, for example, making great
efforts to seek out some supposed herbal or mineral cure, I
have to say: I wonder; have these superstitious chimps *duped*
themselves into relying on a placebo, just as we humans might
have done?

Nevertheless, it's probably only with modern medical
opportunities and new forms of propaganda that the possibility of people being snared by placebos has become a more
serious danger: the danger being that they will indeed indulge
in *premature cure* and the *unwise allocation of resources*. And
it's fortunate perhaps that this danger has increased only at the
same time as, today, people's increased security and affluence,
along with modern regimens of care, have made it actually less
important than it was in the past that they let natural defences,
such as pain, run their course or that they husband resources
for the lean times that might lie ahead.

All the same, the risk is there, and here is another reason to
take it seriously.

It's obviously of some significance that placebo effects are
generally *rewarding* to the subject. To coin a phrase, *placebos
are pleasing*! People prefer to feel good—prefer not to have a
headache or fever or to remain sick—even in those relatively
rare but crucial cases *when continuing to feel bad would be
the safer option*.

Hence, we should hardly be surprised to find that people
continually go looking for placebo treatments, seeking to dis-

cover new forms of placebo—even when they might, as it happens, be better off staying as they are.

But to discover a new placebo, all you need do is to *invent* it, and to invent it all you need do is *change your beliefs*. So it seems the way might well be open for everyone to take *voluntary* and *possibly irresponsible* control of their own health.

Yet, the truth is that—fortunately, perhaps—it's not that easy. When it comes to it, how *do* you change your own beliefs to suit yourself?

No one can simply bootstrap themselves into believing what they choose. Many philosophers, from Pascal through Nietzsche to Orwell, have made this point. But the physicist Stephen Weinberg puts it as nicely as anyone at the end of his book *Dreams of a Final Theory*:

> The decision to believe or not believe is not entirely in our hands. I might be happier and have better manners if I thought I were descended from the emperors of China, but no effort of will on my part can make me believe it, any more than I can will my heart to stop beating.[22]

So what ways—if any—are actually open for the person who longs to believe? Suppose you would desperately like it to be true that, for example, snake oil will relieve your back pain. Which of the belief-creation processes we looked at earlier might you turn to?

Personal experience? Other things being equal, there's little likelihood that snake oil will have worked for you in the past.

Rational argument? There's not going to be much you can do to make a reasoned case for it.

External authority? Ah. Maybe here is the potential weak spot. For surely there will always be somebody out there in the wide world, somebody you find plausible, who will be only too ready to *tell* you that snake oil is the perfect remedy for your bad back.

In the end, then, it's not so surprising that, as we noted at the start of this discussion, people have come to require *outside permission* to get on with the process of self-cure. They may, for sure, have to go looking for it. But who doubts that, in the kind of rich and willing culture we live in, there will be someone, somewhere, to supply them. And around the world and across the ages people have indeed gone looking—seeking the shaman, therapist, guru, or other charismatic healer who can be counted on to tell them the lies they need to believe to make themselves feel good.

However, even as we count the blessings that flow from this tradition of self-deception and delusion, I think that as evolutionary biologists we should keep a critical perspective on it. The fact is it may sometimes be imprudent and improvident to feel this good. Your back pain gets better with the snake oil. But your back pain was designed by natural selection to protect you. And the danger is that while your pain gets better your back gets worse.

Don't tell this to my daughter Ada, or she'll never trust her Dad again.

POSTSCRIPT

Now, friends have told me this isn't the right note to end on. So downbeat. Arguing that placebos are a evolutionary mistake. What about that woman with her metastatic cancer, for example? It seems she owed her life to the snake oil—or rather, in her case, to the guided visualisation. Isn't there a case to be made for the *real health benefits* of submitting to at least some of these illusions?

Yes, happily, there is. And I can and will turn the argument around before I end.

As I've just said, it may sometimes be imprudent for you to show a placebo response when it is unjustified by the reality of your condition. To do so must be counted a biological mistake. But of course you don't want to err in the other direction either. For it could be equally imprudent, or worse, if you were *not* to show a placebo response when it *is* justified by the reality of your condition.

I suspect that just such a situation can arise. And in fact a particularly serious case of it can arise for a surprising reason: which is that your evolved health-care management system may sometimes make *egregious errors* in the allocation of resources—errors which you can only undo by *overriding* the system with a placebo response based on *invalid hope*.

Let me explain. I think we should look more closely at that woman with her cancer. I'd be the last to pretend we really know what was happening in such an extraordinary case. Nonetheless, in the spirit of this essay, I can now make a suggestion. The fact is, or at any rate this the simplest reading of it, that, while the cancer was developing, instead of mounting a full-scale immune response, she continued to withhold some of her immune resources. Earlier in this essay I called this paradoxical: 'Why did she not act before?' But now we have the makings of an explanation: namely that, strange to say, this was quite likely *a calculated piece of health-care budgeting*. Lacking hope for a speedy recovery, the woman was following the rule that says: when in doubt, play safe and keep something in reserve.

Yet, obviously, for her to follow this rule, given the reality of her situation, was to invite potential disaster. To adopt this safety-first policy with regard to future needs was to make it certain there would be no future needs. To withhold her own immune resources as if she might need them in six months' time was to ensure she would be dead within weeks. And so it would have proved, if in fact she had not rebelled in time.

Now, I presume we should not conclude that, because this rule was working out so badly for this woman in this one case, natural selection must have gone wrong in selecting it. Rather, we should recognize that natural selection, in designing human beings to follow these general rules, has had to judge things statistically, on the basis of actuarial data. From natural selection's point of view, this woman would have been a single statistic, whose death as a result of her overcautious health policy would have been more than compensated for by the longer life of her kin, for whom the same policy in somewhat different situations did pay off.

Still, no one could expect the woman herself to have seen things this way. She of course would have wanted to judge things not statistically but individually. And if she could better the built-in general rules by going her own way, using a more up-to-date and relevant model of her personal prospects, this was clearly what she ought to do. Indeed, this ties directly to the argument we've already made about the advantage of fine-tuning the health-care budget on the basis of personal expectancies. It's the very reason why the placebo effect, based on personal hopes, exists. Or, at any rate, why *justified* placebo responses exist.

But in this woman's case, could any placebo response have possibly been *justified*? She certainly had few if any valid grounds for hope. And there is no denying the force of the calculation that she would be taking a big risk with her future health if she were to use up her entire immune resources in one all-or-none onslaught on her cancer, leaving her dangerously vulnerable to further threats down the road.

'Certainly' and 'no denying'—except for one simple fact that changes everything: it's never too early to act when otherwise it's the last act you'll ever have the chance to make. When the alternative is oblivion, hope is always justified. Which is, I think we can say, what she herself proved in real life—when she took matters into her own hands, found her own invented reasons for hoping for a better outcome, released her immune system, beat the cancer, survived the aftermath, and lived on.

Now this is, of course, a special and extreme case. But, in conclusion, I think the reason to be more generally positive about placebos is that several of the same considerations will apply in other less dramatic situations. Your evolved health-care system will have over-erred on the side of caution, and as a result you'll be in unnecessary trouble. It will be appropriate to rebel. And yet having no conventional valid grounds for hope, you too will need to go looking for those rent-a-placebo cures.

Actually, now I come to think about it, the pain of a nettle rash is just such a case of an over-reaction to a rather un-threatening sting.

You can tell this to my daughter Ada, and she'll see that after all her father does know best.

SEDUCTIONS

What Shall We Tell the Children?

'Sticks and stones may break my bones, but words will never hurt me', the proverb goes. And since, like most proverbs, this one captures at least part of the truth, it makes sense that Amnesty International should have devoted most of its efforts to protecting people from the menace of sticks and stones, not words. Worrying about words must have seemed something of a luxury.

Still, the proverb, like most proverbs, is also in part obviously false. The fact is that words *can* hurt. For a start, they can hurt people indirectly by inciting others to hurt them: a crusade preached by a pope, racist propaganda from the Nazis, malevolent gossip from a rival . . . They can hurt people not so indirectly, by inciting them to take actions that harm themselves: the lies of a false prophet, the blackmail of a bully, the flattery of a seducer . . . And words can hurt directly, too: the lash of a malicious tongue, the dreaded message carried by a telegram, the spiteful onslaught that makes the hearer beg his tormentor say no more.

Sometimes, indeed, mere words can kill outright. There is a story by Christopher Cherniak about a deadly 'word-virus' that appeared one night on a computer screen.[1] It took the form of a brain-teaser, a riddle, so paradoxical that it fatally twisted the mind of anyone who heard or read it, making him fall into an irreversible coma. A fiction? Yes, of course. But a fiction with some horrible parallels in the real world. There have been all too many examples historically of how words can take possession of a person's mind, destroying his will to live. Think, for example, of so-called voodoo death. The witch-doctor has merely to cast his spell of death upon a man

and within hours the victim will collapse and die. Or, on a
larger and more dreadful scale, think of the mass suicide at
Jonestown in Guyana in 1972. The cult leader Jim Jones had
only to plant certain crazed ideas in the heads of his disciples,
and at his signal nine hundred of them willingly drank
cyanide.

'Words will never hurt me'? The truth may rather be that
words have a unique power to hurt. And if we were to make
an inventory of the man-made causes of human misery, it
would be words, not sticks and stones, that head the list. Even
guns and high explosives might be considered playthings by
comparison. Vladimir Mayakovsky wrote in his poem 'I':

> On the pavement
> of my trampled soul
> the soles of madmen
> stamp the print of rude, crude, words.[2]

Should we then be fighting Amnesty's battle on this front too?
Should we be campaigning for the rights of human beings
to be protected from verbal oppression and manipulation?
Do we need 'word laws', just as all civilized societies have gun
laws, licensing who should be allowed to use them in what
circumstances? Should there be Geneva protocols establish-
ing what kinds of speech act count as crimes against human-
ity?

No. The answer, I'm sure, ought in general to be 'No, don't
even think of it'. Freedom of speech is too precious a freedom
to be meddled with. And however painful some of its consequences may sometimes be for some people, we should still as
a matter of principle resist putting curbs on it. By all means we
should try to make up for the harm that other people's words
do, but not by censoring the words as such.

And, since I am so sure of this in general, and since I'd
expect most of you to be so too, I shall probably shock you
when I say it is the purpose of this essay to argue in one particular area just the opposite. To argue, in short, in favour of

censorship, against freedom of expression, and to do so, moreover, in an area of life that has traditionally been regarded as sacrosanct.

I am talking about moral and religious education. And especially the education a child receives at home, where parents are allowed—even expected—to determine for their children what counts as truth and falsehood, right and wrong.

Children, I'll argue, have a human right not to have their minds crippled by exposure to other people's bad ideas—no matter *who* these other people are. Parents, correspondingly, have no god-given licence to enculturate their children in whatever ways they personally choose: no right to limit the horizons of their children's knowledge, to bring them up in an atmosphere of dogma and superstition, or to insist they follow the straight and narrow paths of their own faith.

In short, children have a right not to have their minds addled by nonsense. And we as a society have a duty to protect them from it. So we should no more allow parents to teach their children to believe, for example, in the literal truth of the Bible, or that the planets rule their lives, than we should allow parents to knock their children's teeth out or lock them in a dungeon.

That's the negative side of what I want to say. But there will be a positive side as well. If children have a right to be protected from false ideas, they have too a right to be succoured by the truth. And we as a society have a duty to provide it. Therefore we should feel as much obliged to pass on to our children the best scientific and philosophical understanding of the natural world—to teach, for example, the truths of evolution and cosmology, or the methods of rational analysis—as we already feel obliged to feed and shelter them.

I don't suppose you'll doubt my good intentions here. Even so, I realize there may be many readers—especially the more liberal ones—who do not like the sound of this at all: neither the negative nor, still less, the positive side of it.

In which case, among the good questions you may have for me, will probably be these.

First, what is all this about 'truths' and 'lies'? How could anyone these days have the face to argue that the modern scientific view of the world is the only *true* view that there is? Haven't the postmodernists and relativists taught us that more or less anything can be true in its own way? What possible justification could there be, then, for presuming to protect children from one set of ideas or to steer them towards another, if in the end all are all equally valid?

Second, even supposing that in some boring sense the scientific view really is 'more true' than some others, who's to say that this truer world view is the *better* one? At any rate, the better for everybody? Isn't it possible—or actually likely— that particular individuals, given who they are and what their life situation is, would be better served by one of the not-so-true world views? How could it possibly be right to insist on teaching children to think this modern way when, in practice, the more traditional way of thinking might actually work best for them?

Third, even in the unlikely event that almost everybody really would be happier and better off if they were brought up with the modern scientific picture, do we—as a global community—really want everyone right across the world thinking the same way, everyone living in a dreary scientific monoculture? Don't we want pluralism and cultural diversity? A hundred flowers blooming, a hundred schools of thought contending?

And then, last, why—when it comes to it—should *children*'s rights be considered so much more important than those of other people? Everyone would grant of course that children are relatively innocent and relatively vulnerable, and so may have more need of protection than their seniors do. Still, why should their special rights in this respect take precedence over everybody else's rights in other respects? Don't parents have their own rights too, their rights *as* parents? The right, most obviously, to be parents, or literally to bring forth and *pre-pare* their children for the future as *they* see fit?

Good questions? Knock-down questions, some of you may think, and questions to which any broad-minded and progressive person could give only one answer.

I agree they are good-ish questions, and ones that I should deal with. But I don't think it is by any means so obvious what the answers are. Especially for a liberal. Indeed, were we to change the context not so greatly, most people's liberal instincts would, I'm sure, pull quite the other way.

Let's suppose we were talking not about children's minds but children's bodies. Suppose the issue were not who should control a child's intellectual development, but who should control the development of her hands or feet . . . or genitalia. Let's suppose, indeed, that this is an essay about female circumcision. And the issue is not whether anyone should be permitted to deny a girl knowledge of Darwin, but whether anyone should be permitted to deny her the uses of a clitoris.

And now here I am suggesting that it is a girl's right to be left intact, that parents have no right to mutilate their daughters to suit their own socio-sexual agenda, and that we as a society ought to prevent it. What's more, to make the positive case as well, that every girl should actually be encouraged to find out how best to use to her own advantage the intact body she was born with.

Would you still have those good questions for me? And would it still be so obvious what the liberal answers are? There will be a lesson—even if an awful one—in hearing how the questions sound.

First, what's all this about 'intactness' and 'mutilation'? Haven't the anthropological relativists taught us that the idea of there being any such thing as 'absolute intactness' is an illusion, and that girls are—in a way—just as intact without their clitorises?

Anyway, even if uncircumcised girls are in some boring

sense 'more intact', who's to say that intactness is a virtue? Isn't it possible that some girls, given their life situation, would actually be better off being not-so-intact? What if the men of their culture consider intact women unmarriageable?

Besides, who wants to live in a world where all women have standard genitalia? Isn't it essential to maintaining the rich tapestry of human culture that there should be at least a few groups where circumcision is still practised? Doesn't it indirectly enrich the lives of all of us to know that some women somewhere have had their clitorises removed?

In any case, why should it be only the rights of the girls that concern us? Don't other people have rights in relation to circumcision also? How about the rights of the circumcisers themselves, their rights *as* circumcisers? Or the rights of mothers to do what they think best, just as in their day was done to them?

You'll agree, I hope, that the answers go the other way now. But maybe some of you are going to say that this is not playing fair. Whatever the superficial similarities between doing things to a child's body and doing things to her mind, there are also several obvious and important differences. For one thing, the effects of circumcision are final and irreversible, while the effects of even the most restrictive regime of education can perhaps be undone later. For another, circumcision involves the removal of something that is already part of the body and will naturally be missed, while education involves selectively adding new things to the mind that would otherwise never have been there. To be deprived of the pleasures of bodily sensation is an insult on the most personal of levels, but to be deprived of a way of thinking is perhaps no great personal loss.

So, you might argue, the analogy is far too crude for us to learn from it. And those original questions about the rights to control a child's education still need addressing and answering on their own terms.

Very well. I'll try to answer them just so—and we shall see

whether or not the analogy with circumcision was unfair. But there may be another kind of objection to my project that I should deal with first. For it might be argued, I suppose, that the whole issue of intellectual rights is not worth bothering with, since so few of the world's children are in point of fact at risk of being hurt by any *severely* misleading forms of education—and those who are, are mostly far away and out of reach.

Now that I say it, however, I wonder whether anyone could make such a claim with a straight face. Look around—close to home. We ourselves live in a society where most adults—not just a few crazies, but most adults—subscribe to a whole variety of weird and nonsensical beliefs, that in one way or another they shamelessly impose upon their children.

In the United States, for example, it sometimes seems that almost everyone is either a religious fundamentalist or a New Age mystic or both. And even those who aren't will scarcely dare admit it. Opinion polls confirm that, for example, a full 98 per cent of the US population say they believe in God, 70 per cent believe in life after death, 50 per cent believe in human psychic powers, 30 per cent think their lives are directly influenced by the position of the stars (and 70 per cent follow their horoscopes anyway—just in case), and 20 per cent believe they are at risk of being abducted by aliens.[3]

The problem—I mean the problem for children's education—is not just that so many adults positively believe in things that flatly contradict the modern scientific world view, but that so many do not believe in things that are absolutely central to the scientific view. A survey published last year showed that half the American people do not know, for example, that the Earth goes round the Sun once a year. Fewer than one in ten know what a molecule is. More than half do not accept that human beings have evolved from animal ancestors; and less than one in ten believe that evolution—if it has occurred—can have taken place without some kind of external intervention. Not only do people not know the results of science, they do not even know what science is. When asked what they think distinguishes the scientific method, only 2 per cent realized it involves putting theories to the test, 34 per cent vaguely knew it has something to do with experiments and measurement, but 66 per cent didn't have a clue.[4]

Nor do these figures, worrying as they are, give the full picture of what children are up against. They tell us about the beliefs of typical people, and so about the belief environment of the average child. But there are small but significant communities just down the road from us—in New York, or London, or Oxford—where the situation is arguably very much worse: communities where not only are superstition and ignorance even more firmly entrenched, but where this goes hand in hand with the imposition of repressive regimes of social and interpersonal conduct—in relation to hygiene, diet, dress, sex, gender roles, marriage arrangements, and so on. I think, for example, of the Amish Christians, Hasidic Jews, Jehovah's Witnesses, Orthodox Muslims, or, for that matter, the radical New Agers; all no doubt very different from each other, all with their own peculiar hang-ups and neuroses, but alike in providing an intellectual and cultural dungeon for those who live among them.

In theory, maybe, the children need not suffer. Adults might perhaps keep their beliefs to themselves and not make any active attempt to pass them on. But we know, I'm sure, better than to expect that. This kind of self-restraint is simply not in the nature of a parent–child relationship. If a mother, for example, sincerely believes that eating pork is a sin, or that the best cure for depression is holding a crystal to her head, or that after she dies she will be reincarnated as a mongoose, or that Capricorns and Aries are bound to quarrel, she is hardly likely to withhold such serious matters from her own offspring.

But, more important, as Richard Dawkins has explained so well,[5] this kind of self- restraint is not in the nature of successful belief systems. Belief systems in general flourish or die out according to how good they are at reproduction and competition. The better a system is at creating copies of itself, and the better at keeping other rival belief systems at bay, the greater its own chances of evolving and holding its own. So we should expect that it will be characteristic of successful belief systems—especially those that survive when everything else seems to be against them—that their devotees will be obsessed with education and with discipline: insisting on the rightness of their own ways and rubbishing or preventing access to others. We should expect, moreover, that they will make a special point of targeting children in the home, while they are still available, impressionable, and vulnerable. For, as the Jesuit master wisely noted, 'If I have the teaching of children up to seven years of age or thereabouts, I care not who has them afterwards, they are mine for life'.[6]

Donald Kraybill, an anthropologist who made a close study of an Amish community in Pennsylvania, was well placed to observe how this works out in practice:

Groups threatened by cultural extinction must indoctrinate their offspring if they want to preserve their unique heritage. Socialization of the very young is one of the most potent forms of social control. As cultural values slip into the child's mind, they become personal values—embedded in conscience and governed by emotions . . . The Amish contend that the Bible commissions parents to train their children in religious matters as well as the Amish way of life . . . An ethnic nursery, staffed by extended family and church members, moulds the Amish world view in the child's mind from the earliest moments of consciousness.[7]

But what he is describing is not, of course, peculiar to the Amish. 'An ethnic nursery, staffed by extended family and church members' could be as much a description of the early environment of a Belfast Catholic, a Birmingham Sikh, a Brooklyn Hasidic Jew—or maybe the child of a North Oxford don. All sects that are serious about their own survival do indeed make every attempt to flood the child's mind with their own propaganda, and to deny the child access to any *alternative* viewpoints.

In the United States, this kind of restricted education has continually received the blessing of the law. Parents have the legal
right, if they wish, to educate their children entirely at home,
and nearly one million families do so.[8] But many more who
wish to limit what their children learn can rely on the thousands of sectarian schools that are permitted to function subject to only minimal state supervision. A US court did recently
insist that teachers at a Baptist school should at least hold
teaching certificates; but at the same time it recognized that
'the whole purpose of such a school is to foster the development of their children's minds in a religious environment' and
therefore that the school should be allowed to teach all subjects 'in its own way'—which meant, as it happened, present-
ing all subjects only from a biblical point of view, and
requiring all teachers, supervisors, and assistants to agree with
the church's doctrinal position.[9]

Yet, parents hardly need the support of the law to achieve
such a baleful hegemony over their children's minds. For there
are, unfortunately, many ways of isolating children from
external influences without actually physically removing them
or controlling what they hear in class. Dress a little boy in the
uniform of the Hasidim, curl his side-locks, subject him to
strange dietary taboos, make him spend all weekend reading
the Torah, tell him that Gentiles are dirty, and you could send
him to any school in the world and he'd still be a child of the
Hasidim. The same—just change the terms a bit—for a child
of the Muslims, or the Roman Catholics, or followers of the
Maharishi Yogi.

More worrying still, the children themselves may often be
unwitting collaborators in this game of isolation. For children
all too easily learn who they are, what is allowed for them and
where they must not go—even in thought. John Schumaker,
an Australian psychologist, has described his own Catholic
boyhood:

I believed wholeheartedly that I would burn in eternal fire if I ate
meat on a Friday. I now hear that people no longer spend an eternity

in fire for eating meat on Fridays. Yet, I cannot help thinking back on the many Saturdays when I rushed to confess about the bologna and ketchup sandwich I could not resist the day before. I usually hoped I would not die before getting to the 3 p.m. confession.[10]

All the same, this particular Catholic boy actually escaped and lived to tell the tale. In fact Schumaker became an atheist, and has gone on to make something of a profession of his godlessness. Nor, of course, is he unique. There are plenty of other examples, known to all of us, of men and women who as children were pressured into becoming junior members of a sect, Christian, Jewish, Muslim, Marxist—and yet who came out on the other side, free thinkers, and seemingly none the worse for their experience.

Then perhaps I am, after all, being too alarmist about what all this means. For sure, the *risks* are real enough. We do live— even in our advanced, democratic, Western nations—in an environment of spiritual oppression, where many little children—our neighbours' children, if not actually ours—are daily exposed to the attempts of adults to annexe their minds. Yet, you may still want to point out that there's a big difference between what the adults want and what actually transpires. All right, so children do frequently get saddled with adult nonsense. But so what. Maybe it's just something the child has to put up with until he or she is able to leave home and learn better. In which case, I would have to admit that the issue is certainly nothing like so serious as I have been making out. After all, there are surely lots of things that are done to children either accidentally or by design that—though they may not be ideal for the child at the time—have no lasting ill effects.

I'd reply: Yes and No. Yes, it's right we should not fall into the error of a previous era of psychology of assuming that people's values and beliefs are determined once and for all by what they learn—or do not learn—as children. The first years of life, though certainly formative, are not necessarily the 'critical period' they were once thought to be. Psychologists no longer generally believe that children 'imprint' on the first ideas they come across, and thereafter refuse to follow any others. In most cases, rather, it seems that individuals can and will remain open to new opportunities of learning later in life— and, if need be, will be able to make up a surprising amount of lost ground in areas where they have earlier been deprived or been misled.[11]

Yes, I agree therefore we should not be *too* alarmist—or too prissy—about the effects of early learning. But, No, we should certainly not be too sanguine about it, either. True, it may not be so difficult for a person to unlearn or replace factual knowledge later in life: someone who once thought the world was flat, for example, may, when faced by overwhelming evidence to the contrary, grudgingly come round to accepting that the world is round. It will, however, often be very much more difficult for a person to unlearn established procedures or habits of thought: someone who has grown used, for example, to taking everything on trust from biblical authority may find it very hard indeed to adopt a more critical and questioning attitude. And it may be nigh impossible for a person to unlearn attitudes and emotional reactions: someone who has learned as a child, for example, to think of sex as sinful may never again be able to be relaxed about making love.

But there is another even more pressing reason not to be too sanguine, or sanguine in the least. Research has shown that *given the opportunity* individuals can go on learning and can recover from poor childhood environments. However, what we should be worrying about are precisely those cases where such opportunities do not—indeed are not allowed to—occur.

Suppose, as I began to describe above, children are in effect locked out by their families from access to any alternative ideas. Or, worse, that they are so effectively immunized against foreign influences that they do the locking out themselves.

Think of those cases, not so uncommon, when it has become a central plank of someone's belief system that they must not let themselves be defiled by mixing with others.

When, because of their faith, all they want to hear is one voice, and all they want to read is one text. When they treat new ideas as if they carry infection. When, later, as they grow more sophisticated, they come to deride reason as an instrument of Satan. When they regard the humility of unquestioning obedience as a virtue. When they identify ignorance of worldly affairs with spiritual grace . . . In such case, it hardly matters what their minds may still remain capable of learning, because they themselves will have made certain they never again use this capacity.

The question was, does childhood indoctrination matter? And the answer, I regret to say, is that it matters more than you might guess. The Jesuit did know what he was saying. Though human beings are remarkably resilient, the truth is that the effects of *well-designed* indoctrination may still prove irreversible, because one of the effects of such indoctrination will be precisely to remove the means and the motivation to reverse it. Several of these belief systems simply could not survive in a free and open market of comparison and criticism: but they have cunningly seen to it that they don't have to, by enlisting believers as their own gaolers. So, the bright young lad, full of hope and joy and inquisitiveness, becomes in time the nodding elder buried in the Torah; the little maid, fresh to the morning of the world, becomes the washed-up New Age earth mother lost in mists of superstition.

Yet, we can ask, if this is right: what would happen if this kind of vicious circle were to be forcibly broken? What would happen if, for example, there were to be an externally imposed 'time out'? Wouldn't we predict that, just to the extent it *is* a vicious circle, the process of becoming a fully fledged believer might be surprisingly easy to disrupt? I think the clearest evidence of how these belief systems typically hold sway over their followers can in fact be found in historical examples of what has happened when group members have been involuntarily exposed to the fresh air of the outside world.

A interesting test was provided in the 1960s by the case of

the Amish and the military Draft.[12] The Amish have consistently refused to serve in the armed forces of the United States on grounds of conscience. Up to 1960s, young Amish men who were due to be drafted for military service were regularly granted 'agricultural deferments' and were able to continue working safely on their family farms. But as the draft continued through the Vietnam war, an increasing number of these men were deemed ineligible for farm deferments and were required instead to serve two years working in public hospitals—where they were introduced, like it or not, to all manner of non-Amish people and non-Amish ways. Now, when the time came for these men to return home, many no longer wanted to do so and opted to defect. They had tasted the sweets of a more open, adventurous, free-thinking way of life—and they were not about to declare it all a snare and a delusion.

These defections were rightly regarded by Amish leaders as such a serious threat to their culture's survival that they quickly moved to negotiate a special agreement with the government, under which all their draftees could in future be sent to Amish-run farms—so that this kind of breach of security should not happen again.

Let me take stock. I have been discussing the survival strategies of some of the more tenacious beliefs systems—the epidemiology, if you like, of those religions and pseudo-religions that Richard Dawkins has called 'cultural viruses'.[13] But you'll see that, especially with this last example, I have begun to approach the next and more important of the issues I wanted to address: the ethical one.

Suppose that, as the Amish case suggests, young members of such a faith would—if given the opportunity to make up their own minds—*choose to leave*. Doesn't this say something important about the morality of imposing any such faith on children to begin with? I think it does. In fact, I think it says everything we need to know in order to condemn it.

You'll agree that, if it were female circumcision we were

talking about, we could build a moral case against it based just on considering whether it is something a woman would choose for herself. Given the fact—I assume it is a fact—that most of those women who were circumcised as children, if they only knew what they were missing, would have preferred to remain intact. Given that almost no woman who was not circumcised as a child volunteers to undergo the operation later in life. Given, in short, that it seems not to be what free women want to have done to their bodies. Then it seems clear that whoever takes advantage of their temporary power over a child's body to perform the operation must be abusing this power and acting wrongly.

Well, then, if this is so for bodies, it is the same for minds. Given, let's say, that most people who have been brought up as members of a sect, if they only knew what they were being denied, would have preferred to remain outside it. Given that almost no one who was not brought up this way volunteers to adopt the faith later in life. Given, in short, that it is not a faith that a free thinker would adopt. Then, likewise, it seems clear that whoever takes advantage of their temporary power over a child's mind to impose this faith, is equally abusing this power and acting wrongly.

So I'll come to the main point—and lesson—of this essay. I want to propose a general test for deciding when and whether the teaching of a belief system to children is morally defensible, as follows. If it is ever the case that teaching this system to children will mean that later in life they come to hold beliefs that, were they in fact to have had access to alternatives, they would most likely *not* have chosen for themselves, then it is morally wrong of whoever presumes to impose this system and *to chose for them* to do so. No one has the right to *choose badly* for anyone else.

This test, I admit, will not be simple to apply. It is rare enough for there to be the kind of social experiment that occurred with the Amish and the military draft. And even such an experiment does not actually provide so strong a test as I'm suggesting we require. After all, the young Amish men were not offered the alternative until they were already almost grown up, whereas what we need to know is what the children of the Amish or any other sect would choose for themselves if they were to have had access to the full range of alternatives all along. But in practice, of course, such a totally free choice is never going to be available.

Still, utopian as the criterion is, I think its moral implications remain pretty obvious. For, even supposing we cannot know—and can only guess on the basis of weaker tests—whether an individual exercising this genuinely free choice would himself choose the beliefs that others intend to impose upon him, then this state of ignorance in itself must be grounds for making it morally wrong to proceed. In fact, perhaps the best way of expressing this is to put it the other way round, and say: *only* if we know that teaching a system to children will mean that later in life they come to hold beliefs that, were they to have had access to alternatives, they would *still* have chosen for themselves, only then can it be morally allowable for whoever imposes this system and choses for them to do so. And in all other cases, the moral imperative must be to hold off.

Now, I expect most of you will probably be happy to agree with this—so far as it goes. Of course, other things being equal, everybody has a right to self-determination of both body and mind—and it must indeed be morally wrong of others to stand in the way of it. But this is: *other things being equal*. And, to continue with those questions I raised earlier, what happens when other things are not equal?

It is surely a commonplace in ethics that sometimes the rights of individuals have to be limited or even overruled in the interests of the larger good or to protect the rights of other people. And it's certainly not immediately obvious why the case of children's intellectual rights should be an exception.

As we saw, there are several factors that might be considered counterbalancing. And of these, the one that seems to

many people weightiest, or at least the one that is often mentioned first, is our interest as a society in maintaining cultural
diversity. All right, you may want to say, so it's tough on a
child of the Amish, or the Hasidim, or the gypsies to be shaped
up by their parents in the ways they are—but at least the result
is that these fascinating cultural traditions continue. Would
not our whole civilization be impoverished if they were to go?
It's a shame, maybe, when individuals have to be sacrificed to
maintain such diversity. But there it is: it's the price we pay as
a society.

Except, I would feel bound to remind you, *we* do not pay it,
they do.

Let me give a telling example. In 1995, in the high moun-
tains of Peru, some climbers came across the frozen mummi-
fied body of a young Inca girl. She was dressed as a princess.
She was thirteen years old. About five hundred years ago, this
little girl had, it seems, been taken alive up the mountain by a
party of priests, and then ritually killed—a sacrifice to the
mountain's gods in the hope that they would look kindly on
the people below.

The discovery was described by the anthropologist, Johan
Reinhard, in an article for *National Geographic*.[14] He was
clearly elated both as a scientist and as a human being by the
romance of finding this 'ice maiden', as he called her. Even so,
he did express some reservations about how she had come to
be there: 'we can't help but shudder,' he wrote, 'at [the Incas']
practice of performing human sacrifice.'

The discovery was also made the subject of a documentary
film shown on American television. Here, however, no one
expressed any reservation whatsoever. Instead, viewers were
simply invited to marvel at the spiritual commitment of the
Inca priests and to share with the girl on her last journey her
pride and excitement at having been selected for the signal
honour of being sacrificed. The message of the television programme was in effect that the practice of human sacrifice was
in its own way a glorious cultural invention—another jewel in
the crown of multiculturalism, if you like.

Yet, how dare anyone even suggest this? How dare they invite us—in our sitting rooms, watching television—to feel uplifted by contemplating an act of ritual murder: the murder of a dependent child by a group of stupid, puffed up, super-stitious, ignorant old men? How dare they invite us to find good for ourselves in contemplating an immoral action against someone else?

Immoral? By Inca standards? No, that's not what matters. Immoral by ours—and in particular by just the standard of free choice that I was enunciating earlier. The plain fact is that none of *us*, knowing what we do about the way the world works, would freely choose to be sacrificed as she was. And however 'proud' the Inca girl may or may not have been to have had the choice made for her by her family (and for all we know she may actually have felt betrayed and terrified), we can still be pretty sure that she, if she had known what we now know, would not have chosen this fate for herself either.

No, this girl was used by others as a means for achieving their ends. The elders of her community valued their collective security above her life, and decided for her that she must die in order that their crops might grow and they might live. Now, five hundred years later, we ourselves must not, in a lesser way, do the same: by thinking of her death as something that enriches *our* collective culture.

We must not do it here, nor in any other case where we are invited to celebrate other people's subjection to quaint and backward traditions as evidence of what a rich world we live in. We mustn't do it even when it can be argued, as I'd agree it sometimes can be, that the maintenance of these minority traditions is potentially of benefit to all of us because they keep alive ways of thinking that might one day serve as a valuable counterpoint to the majority culture.

The US Supreme Court, in supporting the Amish claim to be exempt from sending their children to public schools, com-mented in an aside: 'We must not forget that in the Middle Ages important values of the civilization of the Western World were preserved by members of religious orders who isolated themselves from all worldly influences against great obstacles.'[15] By analogy, the court implied, we should recognize that the Amish may be preserving ideas and values that our own descendants may one day wish to return to.

But what the court failed to recognize is that there is a crucial difference between the religious communities of the Middle Ages, the monks of Holy Island for example, and the present-day Amish: namely, that the monks made their own choice to become monks, they did not have their monasticism imposed on them as children, and nor did they in their turn impose it on their own children—for indeed they did not have any. Those medieval orders survived by the recruitment of adult volunteers. The Amish, by contrast, survive only by kidnapping little children before they can protest.

The Amish may—possibly—have wonderful things to teach the rest of us; and so may—possibly—the Incas, and so may several other outlying groups. But these things must not be paid for with the children's lives.

This is, surely, the crux of it. It is a cornerstone of every decent moral system, stated explicitly by Immanuel Kant but already implicit in most people's very idea of morality, that human individuals have an absolute right to be treated as *ends in themselves*—and never as *means* to achieving other people's ends. It goes without saying that this right applies no less to children than to anybody else. And since, in so many situations, children are in no position to look after themselves, it is morally obvious that the rest of us have a particular duty to watch out for them.

So, in every case where we come across examples of children's lives being manipulated to serve other ends, we have a duty to protest. And this, no matter whether the other ends involve the mollification of the gods, 'the preservation of important values for Western civilization', the creation of an interesting anthropological exhibit for the rest of us . . . or— now I'll come to the next big question that's been waiting—the fulfilment of certain needs and aspirations of the child's own parents.

There is, I'd say, no reason whatever why we should treat the actions of parents as coming under a different set of moral rules here.

The relationship of parent to child is of course a special one in all sorts of ways. But it is not so special as to deny the child her individual personhood. It is not a relationship of co-extension, nor one of ownership. Children are not a part of their parents, nor except figuratively do they 'belong' to them. Children are in no sense their parents' private property. Indeed, to quote the US Supreme Court, commenting in a different context on this same issue: it is a 'moral fact that a person belongs to himself and not others nor to society as a whole'.[16]

It will therefore be as much a breach of a child's rights if he or she is used by his or her parents to achieve the parents' personal goals, as it would be if this were done by anyone else. No one has a right to treat children as anything less than ends in themselves.

Still, some of you I'm sure will want to argue that the case of parents is not *quite* the same as that of outsiders. No doubt we'd all agree that parents have no more right than anyone else to exploit children for ends that are obviously selfish—to abuse them sexually, for example, or to exploit them as servants, or to sell them into slavery. But, first, isn't it different when the parents at least think their own ends are the child's ends too? When their manipulation of the child's beliefs to conform to theirs is—so as far as they are concerned—entirely in the child's best interests? And then, second, isn't it different when the parents have already invested so much of their own resources in the child, giving him or her so much of their love and care and time? Haven't they somehow earned the reward of having the child honour their beliefs, even if these beliefs are—by other people's lights—eccentric or old-fashioned?

Don't these considerations, together, mean that parents have at least some rights that other people don't have? And rights which arguably should come before—or at least rank beside—the rights of the children themselves?

No. The truth is these considerations simply don't add up to any form of *rights,* let alone rights that could outweigh the children's: at most they merely provide mitigating circumstances. Imagine: suppose you were misguidedly to give your own child poison. The fact that you might think the poison you were administering was good for your child, the fact that you might have gone to a lot of trouble to obtain this poison, and that if it were not for all your efforts your child would not even have been there to be offered it, none of this would give you *a right* to administer the poison—at most, it would only make you less culpable when the child died.

But in any case, to see the parents as simply misguided about the child's true interests is, I think, to put too generous a construction on it. For it is not at all clear that parents, when they take control of their children's spiritual and intellectual lives, really do believe they are acting in the child's best interests rather than their own. Abraham, when he was commanded by God on the mountain to kill his son, Isaac, and dutifully went ahead with the preparation, was surely not thinking of what was best for Isaac—he was thinking of his own relationship with God. And so on down the ages. Parents have used and still use their children to bring *themselves* spiritual or social benefits: dressing them up, educating them, baptising them, bringing them to confirmation or Bar Mitzvah in order to maintain their own social and religious standing.

Consider again the analogy with circumcision. No one should make the mistake of supposing that female circumcision, in those places where it's practised, is done to benefit the girl. Rather, it is done for the honour of the family, to demonstrate the parents' commitment to a tradition, to save them from dishonour. Although I would not push the analogy too far, I think the motivation of the parents is not so different at many other levels of parental manipulation—even when it comes to such apparently unselfish acts as deciding what a child should or should not learn in school.

A Christian fundamentalist mother, for example, forbids her child from attending classes on evolution: though she may claim she is doing it for the child and not of course herself, she is very likely motivated primarily by a desire to make a display of her own purity. Doesn't she just know that God is mighty proud of her for conforming to his will? . . . The chief mullah of Saudi Arabia proclaims that the Earth is flat and that anyone who teaches otherwise is a friend of Satan:[17] won't he himself be thrice blessed by Allah for making this courageous stand? A group of rabbis in Jerusalem try to ban the showing of the film *Jurassic Park* on the grounds that it may give children the idea that there were dinosaurs living on earth sixty million years ago, when the scriptures state that in fact the world is just six thousand years old:[18] are they not making a wonderful public demonstration of their own piety?

What we are seeing, as often as not, is pure self-interest. In which case, we should not even allow a mitigating plea of good intentions on the part of the parent or other responsible adult. They are looking after none other than themselves.

Yet, as I said, in the end it hardly matters what the parents' intentions are; because even the best of intentions would not be sufficient to buy them 'parental rights' over their children. Indeed, the very idea that parents or any other adults have 'rights' *over* children is morally insupportable.

No human being, in any other circumstances, is credited with having rights *over* any one else. No one is entitled, as of right, to control, use, or direct the life-course of another person—even for objectively good ends. It's true that in the past slave-owners had such legal rights over their slaves. And it's true, too, that, until comparatively recently, the anomaly persisted of husbands having certain such rights over their wives—the right to have sex with them, for instance. But neither of these exceptions provides a good model for regulating parent–child relationships.

Children, to repeat, have to be considered as having interests independent of their parents. They cannot be subsumed as if they were part of the same person. At least so it should be. Unless, that is, we make the extraordinary mistake that the US Supreme Court apparently did when it ruled, in relation to the Amish, that while the Amish way of life may be considered 'odd or even erratic', it 'interferes with no rights or interests of *others* (my italics).[19] As if the children of the Amish are not even to be counted as potentially 'others'.

I think we should stop talking of 'parental rights' at all. In so far as they compromise the child's rights as an individual, parents' rights have no status in ethics and should have none in law.[20]

That's not to say that, other things being equal, parents should not be treated by the rest of us with due respect and accorded certain 'privileges' in relation to their children. 'Privileges', however, do not have the same legal or moral significance as rights. Privileges are by no means unconditional; they come as the quid pro quo for agreeing to abide by certain rules of conduct imposed by society at large. And anyone to whom a privilege is granted remains in effect on probation: a privilege granted can be taken away.

Let's suppose that the privilege of parenting will mean, for example, that, provided parents agree to act within a certain framework, they shall indeed be allowed—without interference from the law—to do all the things that parents everywhere usually do: feeding, clothing, educating, disciplining their own children—and enjoying the love and creative involvement that follow from it. But it will explicitly *not* be part of this deal that parents should be allowed to offend against the child's more fundamental rights to self-determination. If parents do abuse their privileges in this regard, the contract lapses—and it is then the duty of those who granted the privilege to intervene.

Intervene *how*? Suppose we—I mean we as a society—do not like what is happening when the education of a child has been left to parents or priests. Suppose we fear for the child's mind and want to take remedial action. Suppose, indeed, we want to take pre-emptive action with *all* children to protect them from being hurt by bad ideas and to give them the best possible start as thoughtful human beings. What should we be doing about it? What should be our birthday present to them from the grown-up world?

My suggestion at the start of this essay was: science—universal scientific education. That's to say, education in learning from observation, experiment, hypothesis testing, constructive doubt, critical thinking—and the truths that flow from it.

And so I've come at last to the most provocative of the questions I began with. What's so special about science? Why *these* truths? Why should it be morally right to teach *this* to everybody, when it's apparently so morally wrong to teach all those other things?

You do not have to be one of those out-and-out relativists to raise such questions—and to be suspicious that any attempt to replace the old truths by newer scientific truths might be nothing other than an attempt to replace one dogmatism by another. The Supreme Court, in its judgement about Amish schooling, was careful to warn that we should never rule out one way of thinking and rule another in, merely on the basis of what happens to be the modern, fashionable opinion. 'There can be no assumption', it said, 'that today's majority is "right" and the Amish and others are "wrong"'; the Amish way of life 'is not to be condemned because it is different'.[21]

Maybe so. And yet I'd say the court has chosen to focus on the wrong issue there. Even if science *were* the 'majority' world view (which, as we saw earlier, is sadly not the case), we'd all agree that this in itself would provide no grounds for promoting science above other systems of thought. The 'majority' is clearly not right about lots of things, probably most things.

But the grounds I'm proposing are firmer. Some of the other speakers in the lecture series in which this essay first had an airing will have talked about the values and virtues of science. And I am sure they too, in their own terms, will have attempted to explain why science is different—why it ought to have a unique claim on our heads and on our hearts. But I will now perhaps go even further than they would. I think science stands apart from and superior to all other systems for the reason that it alone of all the systems in contention meets the criterion I laid out above: namely, that it represents a set of beliefs that any reasonable person would, if given the chance, choose for himself.

I should probably say that again, and put it in context. I argued earlier that the only circumstances under which it should be morally acceptable to impose a particular way of thinking on children is when the result will be that later in life they come to hold beliefs that they would have chosen anyway, no matter what alternative beliefs they were exposed to. And what I am now saying is that science is the one way of thinking—maybe the only one—that passes this test. There is a fundamental asymmetry between science and everything else.

What do you reckon? Let's go to the rescue of that Inca girl who is being told by the priests that, unless she dies on the mountain, the gods will rain down lava on her village, and let's offer her another way of looking at things. Offer her a choice as to how she will grow up: on one side with this story about divine anger, on the other with the insights from geology as to how volcanoes arise from the movement of tectonic plates. Which will she choose to follow?

Let's go help the Muslim boy who's being schooled by the mullahs into believing that the Earth is flat, and let's explore some of the ideas of scientific geography with him. Better still, let's take him up high in a balloon, show him the horizon, and invite him to use his own senses and powers of reasoning to reach his own conclusions. Now, offer him the choice: the picture presented in the book of the Koran, or the one that flows from his new-found scientific understanding. Which will he prefer?

Or let's take pity on the Baptist teacher who has become wedded to creationism, and let's give her a vacation. Let's walk her round the Natural History Museum in the company of Richard Dawkins or Dan Dennett—or, if they're too scary, David Attenborough—and let's have them explain the possibilities of evolution to her. Now, offer her the choice: the story of Genesis with all its paradoxes and special pleading, or the startlingly simple idea of natural selection. Which will she choose?

My questions are rhetorical because the answers are already in. We know very well which way people will go when they really are allowed to make up their own minds on questions such as these. Conversions from superstition to science have been and are everyday events. They have probably been part of our personal experience. Those who have been walking in darkness have seen a great light: the aha! of scientific revelation.

By contrast, conversions from science back to superstition are virtually unknown. It just does not happen that someone who has learnt and understood science and its methods and who is then offered a non-scientific alternative chooses to abandon science. I doubt there has ever been a case, for example, of someone who has been brought up to believe the geological theory of volcanoes moving over to believing in divine anger instead, or of someone who has seen and appreciated the evidence that the world is round reverting to the idea that the world is flat, or even of someone who has once understood the power of Darwinian theory going back to preferring the story of Genesis.

People do, of course, sometimes abandon their existing scientific beliefs in favour of newer and better scientific alternatives. But to choose one scientific theory over another is still to remain absolutely true to science.

The reason for this asymmetry between science and non-science is not—at least not only—that science provides so much better—so much more economical, elegant, beautiful—explanations than non-science, although there is that. The still stronger reason, I'd suggest, is that science is by its very nature a participatory process and non-science is not.

In learning science *we* learn *why we should believe* this or that. Science doesn't cajole, it doesn't dictate, it lays out the factual and theoretical arguments as to why something is so—and invites us to assent to them, to see it for ourselves. Hence, by the time someone has understood a scientific explanation, they have in an important sense already chosen it as theirs.

How different is the case of religious or superstitious explanation. Religion makes no pretence of engaging its devotees in any process of rational discovery or choice. If we dare ask *why* we should believe something, the answer will be because it has been written in the Book, because this is our tradition, because it was good enough for Moses, because you'll go to heaven that way . . . Or, as often as not, don't ask.

Contrast these two positions. On one side, there is the second-century Roman theologian Tertullian, with his abject submission to authority and denial of our personal involvement in choosing our beliefs. 'For us,' he wrote, 'curiosity is no longer necessary after Jesus Christ nor inquiry after the Gospel.'[22] This is the same man, I might remind you, who said of Christianity: 'It is certain because it is impossible.' On the other side, there is the twelfth-century English philosopher, Adelard of Bath, one of the earliest interpreters of Arab science, with his injunction that we all make ourselves personally responsible for understanding what goes on around us. 'If anyone living in a house is ignorant of what it is made . . . he is unworthy of its shelter,' he said, 'and if anyone born in the residence of this world neglects learning the plan underlying its marvellous beauty . . . he is unworthy . . . and deserves to be cast out of it.'[23]

Imagine that the choice is yours: that you have been faced, in the formative years of your life, with a choice between these two paths to enlightenment—between basing your beliefs on

the ideas of others imported from another country and
another time, and basing them on ideas that you have been
able to see growing in your home soil. Can there be any doubt
that you will choose for yourself, that you will choose science?

And because people will so choose, *if* they have the opportunity of scientific education, I say we as a society are entitled
with good conscience to *insist* on their being given that opportunity. That is, we are entitled in effect to choose this way of
thinking for them. Indeed, we are not just entitled: in the case
of children, we are morally obliged to do so—so as to protect
them from being early victims of other ways of thinking that
would remove them from the field.

Then—suppose someone asks—'How would *you* like it if
some Big Brother were to insist on *your* children being taught
his beliefs? How'd you like it if *I* tried to impose my personal
ideology on *your* little girl?' I have the answer: that teaching
science isn't like that. It's not about teaching someone else's
beliefs, it's about encouraging the child to exercise her powers
of understanding to arrive at her own beliefs.

For sure, this is likely to mean she will end up with beliefs
that are widely shared with others who have taken the same
path: beliefs, that is, in what science reveals as the *truth* about
the world. And yes, if you want to put it this way, you could
say this means that by her own efforts at understanding she
will have become a scientific conformist: one of those predictable people who believes that matter is made of atoms,
that the universe arose with the Big Bang, that humans are
descended from monkeys, that consciousness is a function of
the brain, that there is no life after death, and so on . . . But—
since you ask—I'll say I'd be only too pleased if a big brother
or sister or school-teacher or you yourself, sir, should help her
get to that enlightened state.

The habit of questioning, the ability to tell good answers
from bad, an appetite for seeing how and why deep explanations work—such is what I would want for my daughter
(now two years old) because I think it is what she, given the

chance, would one day want for herself. But it is also what I would want for her because I am too well aware of what might otherwise befall her. Bad ideas continue to swill through our culture, some old, some new, looking for receptive minds to capture. If this girl, lacking the defences of critical reasoning, were ever to fall to some kind of political or spiritual irrationalism, then I and you—and our society—would have failed her.

Words? Children are made of the words they hear. It matters what we tell them. They can be hurt by words. They may go on to hurt themselves still further, and in turn become the kind of people that hurt others. But they can be given life by words as well.

'I have set before you life and death, blessing and cursing,' in the words of Deuteronomy, 'therefore choose life, that both thou and thy seed may live.'[24] I think there should be no limit to our duty to help children to choose life.[25]

21

The Number of the Beast

The number 666 is not a number to take lightly. If the biblical book of Revelation is to be relied on, it is the number that at the Day of Judgement will mark us for our doom. 'Here is wisdom. Let him that hath understanding count the number of the beast; for it is the number of a man, and his number is six hundred, three score and six.'[1]

But why, if I dare ask it, precisely 666? Why is the number of the beast also 'the number of a man'? And why, according some other ancient authorities, is the correct number not 666 but 616? I believe a curious passage in Plato's *Republic* holds the answer.

In Book VIII of the *Republic*, Socrates discusses how the Guardians should regulate the supply of newborn children, and he draws attention to a potential problem with meeting quotas: 'Though the Rulers you have trained for your city are wise, reason and perception will not always enable them to hit on the right and wrong times for breeding; some time they will miss them and then children will be begotten amiss.' Fortunately, he says, it can all be worked out by mathematics:

For the number of the human creature is the first in which root and square multiplications (comprising three dimensions and four limits) of basic numbers, which make like and unlike, and which increase and decrease, produce a final result in completely commensurate terms.[2]

The meaning is not as clear as we might wish. However, classical scholars generally agree that the number Socrates is trying to specify here—which has come to be called 'Plato's number'—is the number of days that is the minimum needed for gestation. This number was believed by the Greeks to be 216 days, or seven months (while normal gestation is 276 days, or nine months, which is $216 + 3 \times 4 \times 5$).

Now, 216 has various special properties. To start with, it is six cubed. The number six is supposed to represent marriage and conception, since $6 = 2 \times 3$, and the number two represents femininity and three masculinity. The number six is also the area (a square measure) of the well-known Pythagorean triangle with sides three, four, and five. And six cubed is the smallest cube that is the sum of three other cubes: namely, the cubes of three, four and five. That's to say, $216 = 6^3 = 3^3 + 4^3 + 5^3$.

It seems to have been this last property of the important number that Plato was trying to describe. But since there was no word in classical Greek to express 'cubing', the best he could do was to employ the clumsy expression 'root and square multiplications (comprising three distances and four limits)'.

So, where is the connection to 666, the number of the beast? Note, if you will, the following coincidences:

- Revelation says, 'For it is the number of a man'; Plato says, 'For the number of the human creature is . . .'

- Revelation gives the number as 666; Plato gives it as $6 \times 6 \times 6$.

- The alternative number of the beast is 616; the alternative way of expressing Plato's number is 216.

The *Republic* was written four hundred years before the book of Revelation. John the Divine, the probable author of Revelation, wrote in Greek and, although relatively uneducated, he would very likely have had at least a passing acquaintance with Plato's ideas. I suggest John simply misunderstood Plato's numeric formulation and took the three sixes in the wrong combination. Other authorities then corrected the last two digits, to put back the 16 to make 616.

22

Arms and the Man

Some years ago there was a competition in the *New Statesman* to produce the most startling newspaper headline anyone could think of. Among the winning entries, as I remember, was this: 'Archduke Ferdinand Still Alive: First World War a Mistake'.

A mistake, a war in which twenty million people died? It seems, of course, preposterous: a bad—if clever—joke. And yet historians now almost universally agree that the First World War *was* a mistake. Not in the sense implied by that headline, that it was fought for the *wrong reasons*. No, a mistake in the sense—perhaps even more depressing—that there were no reasons for war at all: *except, that is, for the drive to war itself.* The great powers had got themselves into a fierce arms race, and Germany was being made to feel increasingly encircled; yet rather than seeking obvious remedies— reversing the arms race, or resorting to more positive diplomacy—the generals and the statesmen pushed on towards the cataclysm as if they were no longer masters of their fate.

In recent history there have been, as we know, disturbing parallels. In the late years of the twentieth century, the great civilizations of the world—the very nations that gave us our literature, music, painting, our ideas of peace and of democracy—were once again locked into a unending cycle of hostility. Once again there was a grim competition for superiority in arms, in trade, in attitudes, in righteousness. We called it a Cold War—no one was being killed. But let no one doubt that it was war. Year by year more money was being spent on armaments than at any time during the hot wars of the past.

As a direct consequence of this misuse of resources, thousands of people daily starved.

There were, it is true, lulls in the hostility. Moments of hope. Moments when, against the odds, it seemed that world leaders might be coming to their senses. But soon enough they took up where they left off, as if hurrying to join step in the long march to oblivion.

In *Mother Courage*, Bertolt Brecht wrote of the seemingly unstoppable energy of war. Says Mother Courage:

Well, there've always been people going around saying the war will end. I say, you can't be sure the war will *ever* end. Of course it may have to pause occasionally—for breath, as it were—it can even meet with an accident—nothing on this earth is perfect. A little oversight, and the war's in the hole, and someone's got to pull it out again! That someone is the Emperor or the King or the Pope. They're such friends in need, the war has really nothing to worry about, it can look forward to a prosperous future.[1]

Brecht wrote of war as if it were a kind of monstrous, living force. And, indeed, to many others writing in the twentieth century, it has seemed that war *is* a kind of autonomous, self-perpetuating, and self-serving agency. Do we need reminding, then, that wars are made by human beings?

I hope I may be allowed to talk a bit about matters of psychology. Professional psychologists have not always, it is true, had anything sensible or relevant to say about what has been so obviously the greatest issue of our time. But in recent years new ideas have emerged. And, surprisingly enough, one of the most fertile areas has been 'games theory'.

Games theory, as John von Neumann originally conceived it, is the theory of how rational people should behave in any game where the interests of the players are conflicting. But while games theory deals essentially with rationality, one of the startling results has been to show just how *irrational* rational behaviour may become. For there are situations where, because of factors inherent in the situation itself, sensible players pursuing apparently sensible short-term goals always end up with exactly the opposite result of that which they intended. In psychology, this has led to the notion of what's called a 'social trap'.

Let us take an artificial situation to illustrate the kind of thing I mean. There is a notorious game called the 'pound auction'. Suppose that someone says he is going to sell a £1 coin to the highest bidder. And he starts by asking for bids of 10p. It sounds good value, and no doubt most of us would come in with a bid. But there is a snag. For the auctioneer announces that he is going to take the money not just from the highest bidder but from the next highest as well. Well, that's a shame . . . still, we'll put up with it. So we bid 10p. Someone on the other side comes in with a bid of 20p. Now what are we going to do? If we leave it there we will have lost out. Not only will we not get the £1, we will lose our 10p all for nothing. So we bid 30p. A pound for 30p—still not a bad deal if we can get it. But now the other contestant is in exactly the position we were—he's going to lose out unless he continues. So he ups his bid to 40p. So we up ours again, so he ups his. You might think the auction is bound to stop when either we or the other bidder reaches a full £1. But not at all. For suppose the auction has reached a point where our rival has bid 90p and we have bid £1. Unless he now goes *over* £1, he is going to lose everything, while he can at least cut his losses by bidding £1.10. So we bid £1.20. And so on, onwards and upwards—while both of us get deeper into debt. The fact is that from the moment we entered this game, both sides were trapped.

Now the problem, as you will see, arises precisely because it is built into the situation that our gain is his loss, and vice versa. And the parallel to an arms race between nations is all too obvious. The very weapons which give one side a sense of security threaten the other—and naturally, and even rationally, the other side retaliates in kind.

Yet, surely, you might think, what goes up can come down, and a vicious spiral can in principle become a virtuous one. Why is it then that these traps are so hard to get out of—why cannot the whole thing just be stopped or even put into reverse?

All the evidence suggests that stopping, let alone reversing, is in fact very difficult. The reason is quite simple, namely, that it requires cooperation between two sides who are defined by the very situation as antagonists. If we take the pound auction, it's obvious that if at a stage where one side has bid 10p, the other 20p, both were to stop *and agree to split the profits,* both would be better off. But the problem is exactly that. It requires *agreement* between two participants who by the very nature of the game are in an unbalanced relationship. Someone is always in front: someone always thinks he stands to gain more by winning the race than pulling out of it; what's more, each side can of course blame the other for the costs incurred so far.

Cooperation requires trust and generosity. Many people assume that the chief obstacle to cooperation is hatred of the enemy. But surprisingly, a peculiar feature of the games-playing approach to human conflict is that it does not place any great emphasis on *hatred*. Hatred can, of course, make a conflict less easy to resolve. But hatred is in no way a *necessary* feature of an arms race. Indeed, it is important to realize just how fast and far a conflict can proceed without, as it were, ever an angry word being spoken.

Before the First World War, for example, the Germans did not hate the British, nor the British the Germans—far from it, they were in most respects firm friends; and the hatred that emerged was secondary—a consequence of the war rather than its cause. Similarly in the 1980s, *at least at the level of professional war-planners,* there was as a matter of fact very little outright enmity. Why should there have been? For in a curious sense the American and Soviet war-planners, in spite of being formal adversaries, were in fact allies with a common interest in the theory of deterrence and the problem of how to win a no-win game. The 'toys' they played with were only incidentally weapons designed to *hurt* other human beings—as if that, too, was some kind of unfortunate mistake. 'Hatred', as the American games theorist Anatol Rappoport put it, 'needs to play no role in the war-planners' involvement with the wherewithal of omnicide because it is no longer necessary to hate anyone in order to kill everyone.'

Yet while 'hatred of the enemy' may not be a necessary feature of the drive to war, it is surely a facilitating factor, providing if nothing else a social context in which the war-planners can pursue their deadly games. Hatred, moreover, is a distorting emotion which when present is bound to block any rational analysis of how to achieve genuine security. It is in this area that I believe psychology still has to make its greatest contribution.

Why do people *dislike* other people? The obvious answer, you might think, is that we dislike others who either have or might cause *us* harm. But a surprisingly different answer has come forward. We tend to dislike people not because of anything *they* have done to us, but because of the things *we* have done to them. In other words, hostility and hatred are self-fulfilling: we develop our hostile attitudes as a consequence of our own actions, *we learn by doing*.

If I hurt someone else, you might think that it would make me feel tenderly disposed to him or her. But research, especially on children, shows quite the opposite. A child who hurts another, even by accident, is more likely than not to think the worse of the victim—to invent reasons, in short, for why it served him or her right. And while adults generally have more sophisticated feelings, the same holds true. It is as though people need to rationalize their own actions after the event, and the only way to explain how we have caused harm to someone else is to persuade ourselves that in some way or another they deserved it.

We see it in a burglar who despises the householders he steals from. We saw it frighteningly in Vietnam, where American soldiers developed ever greater hatred for the peasants the more they killed and maimed them. And so all the way through to the behaviour of whole nations: the Nazis loathing the Jews *because* the Jews suffered at their hands, the Israelis hating the Palestinians because they are weak and poor.

More disturbing still, the acts of antagonism that lead to

hatred do not even have to be real acts: they can be purely imaginary. If, say, I merely plan to hurt someone else, the very idea of the pain that I might cause him may lead me to devalue him. What, then, if the hurt I am planning is an act of ultimate destruction—nuclear obliteration? No wonder, perhaps, that Ronald Reagan saw the Russians as the 'focus of evil in the modern world', when every day he—Reagan—must have lived with the idea of *turning* Russia into Hell.

Let us note here one of the many paradoxes of nuclear weapons. No one can doubt any longer that their use would be an act of suicide. When the President of the United States imagined killing Russians he must have imaginde too killing his own countrymen and killing his own self. But when to kill someone even in imagination *leads* to hatred, the man with his finger on the button must slowly have been growing deep down to hate himself.

Yet there is another side to it, another side to this 'learning by doing'. What psychologists have found is that just as hostil- ity can be self-fulfilling, so can friendship, so can love. In fact, perhaps it is not surprising that just as we need to find reasons to explain our own hostile actions, so we need reasons to explain our acts of generosity.

And so it happens that someone who is persuaded, even tricked, into being kind to someone else actually comes, as a result of his own actions, to value that other person more. A child, for example, who is encouraged by his teacher to make toys for poor children in hospital begins to reckon that the recipients deserve his help. The millions of people who gave money for starving Ethiopians in the mid-eighties almost cer- tainly came to care about them as human beings. Does the same principle carry through right up to the behaviour of whole nations? The fact is, we do not know. Why not? *We do not know because it never has been tried.*

I was talking a few years ago to Ervin Staub, a Hungarian psychologist who has been responsible as much as anyone for this idea of 'learning by doing'.[2] I asked him: just suppose that the United States were to have found itself, by accident as it were, being generous to the Russians. Suppose the disaster at Chernobyl had been so overwhelming that the Russians had required the kind of aid that only the United States could give. Is it possible that such an unthought-out act of charity by the United States would have marked the end of the Cold War? Professor Staub's answer was that he had indeed hoped that it would happen: that out of darkness might have come an unexpected light.

I do not often have the privilege of speaking in a church. But when I do I confess that the echo of the place gets to me, and I find myself wanting to quote passages of scripture. So let me remind you now of that ambiguous and somewhat frightening injunction from St Matthew's gospel: 'Unto everyone that hath shall be given and he shall have abundance; but from him that hath not shall be taken away even that which he hath.'[3]

For a long time I did not understand it. But now I think I do. For in that parable of the talents, Jesus was perhaps talking first and foremost about human love: 'Unto everyone that hath love shall be given the power to love, but from him that hath not shall be taken away even the power which he hath.'

Consider the story in this light.

Then he that had received the five talents went and traded with the same, and made them another five talents. And likewise he that had received two he also gained another two. But he that had received one went and digged in the earth and hid his lord's money . . . 'I was afraid,' said the wicked servant, 'and went and hid thy talent in the earth.'[4]

When we find ourselves living in a 'world in peril', it may be because we have hidden our talent for loving each other in the earth.

23

Death in Tripoli

Colonel Ghadaffi's daughter would have been four years old. As the year 1987 broke she might have been enjoying the North African sun, laughing at the sea, wriggling her toes into the sand . . . A shame, then, that she had been killed by a bomb. How much of a shame? It depends on how you calculate it. But shall we say forty-five seconds, at the most?

As with other years, there were many reasons for disliking things said and done by politicians in 1986. But the one that stuck with me was that argument between Mr Norman Tebbit and the BBC about the precise time that ought to have been devoted on the television news to the killing of Ghadaffi's little girl.

In 1984, I, along with most of the British population, saw the pictures of Mr Tebbit himself being dragged in pain from the ruins of a Brighton hotel. No one, so far as I know, protested then that the pictures were being used for propaganda; nor subsequently did anyone complain about the time given to the plight of Mr Tebbit's injured wife. To have done so would have been petty and immoral. Here were pictures that the British public—and the IRA—did well to see. Bombs kill, that was the lesson; and the killing, even of political enemies, is wrong.

If nothing else, a sense of fair dealing might therefore have prevented Mr Tebbit from grudging the public a minute's exposure to the contemplation of someone else's pain. Mr Tebbit, knowing what he did from personal experience, might even have been moved to pity. But no, when Tripoli was bombed, the chairman of the Conservative Party was alarmed for nothing else than his own party's skin. He counted the BBC out, he counted the ITV in; he *timed* the agony—and found the baby's death sixty-one seconds too long.

Henry VI was criticized for having spent too much money on his chapel at King's College, Cambridge. Later, Wordsworth wrote:

> High Heaven rejects the lore
> of nicely calculated less and more.[1]

The words, if not the context for them, apply exactly here. To calculate the costs of a child's life in terms of its political advantage to oneself—or to an enemy—is ethically degenerate. Why? Because it involves a shift in perception that good people ought never to allow: from the child as a being who has value in herself, to the child as a mere term in someone else's sum.

Immanuel Kant, as a philosopher, wrote this:

> When the question is whether a thing be beautiful, we do not want to know whether anything depends or can depend, for us or for anyone else, on the existence of the object, but only how we estimate it in pure contemplation.[2]

And the same is true, he thought, for acts of morality and immorality. There is indeed only one ethical imperative: that we should regard all other human beings as *ends*, and never as *means* to our own ends.

Kant's code was strict; and none of us in the real world live up to it. But when we fail, is there at least some way that we can begin to atone for the immoral act? Psychologically, and arguably philosophically as well, the remedy—such as it is— can lie only in confronting like with like: we must, in short, pay for our neglect of another person's intrinsic worth by a deliberate humbling of our own.

Other cultures knew it. Two and a half thousand years ago, Lao-Tzu wrote:

> Those who have killed their thousands in the fray
> should shed hot tears, and celebrate the day
> with funeral rites such as wan mourners pay.[3]

Shakespeare, in *Richard II*, had Bolingbroke rebuking the killer of King Richard:

> Exton, I thank thee not . . . though I did wish him dead,
> I hate the murderer, love him murdered . . .
> Come, mourn with me for what I do lament,
> and put on sullen black incontinent.[4]

Imagine now a society where the newsreaders on the BBC would—by direction of the government—wear the signs of mourning at each and every act of cruelty initiated or condoned by the UK: a society where the *Sun* would shout 'FORGIVE US!' when we torpedoed the *Belgrano*, and where the Prime Minister on the steps of Downing street would cry 'Lament!'.

It will not happen here. Yet such a society might, I think, come closer to a state of grace. What is more, it would win the propaganda battle, too. For the truth is that, while to act in one's own self-interest cannot be moral, to act morally can often be in one's self-interest. A paradox? Yes, *the* paradox of morals—and one that can only be confirmed, and not explained.

24

Follow My Leader

Ian Kershaw, in his biography of Hitler, quotes a teenage girl, writing to celebrate Hitler's fiftieth birthday in April 1939: 'a great man, a genius, a person sent to us from heaven'.[1] What kind of design flaw in human nature could be responsible for such a seemingly grotesque piece of hero-worship? Why do people in general fall so easily under the sway of dictators?

I shall make a simple suggestion at the end of this essay. But I shall not go straight there. The invitation to talk about 'Dictatorship: The Seduction of the Masses' in a series of seminars on 'History and Human Nature' has set me thinking more generally about the role of evolutionary theory in understanding human history. And I've surprised myself by some of my own thoughts. I'll come to dictatorship in a short while, but I want to establish a particular framework first.

The usual reason that's given for paying attention to evolutionary psychology is that, by exposing and explaining human nature, it can show us the *constraints* that have operated—and still do—on the historical development of human culture. I'll quote, if I may, from the manifesto I myself wrote for the Darwin Centre at the London School of Economics:

Evolutionary theory leads us to expect that, to the extent that human beings have evolved to function optimally under the particular social and ecological conditions that prevailed long ago, they are likely still today to be better suited to certain life-styles and environments than others. What's more they can be expected to bring with them into modern contexts archaic ways of thinking, preferences and prejudices, even when these are no longer adaptive (or even positively maladaptive).

The emphasis, you'll see, is on how nature and culture *conflict*. 'We're Stone Age creatures living in the space age', as the saying goes. And the implication is that the modernizing forces of culture have continually to fight a rearguard battle against the recidivist tendencies of human nature. Almost every evolutionary psychology textbook ends with a sermon on the importance of getting in touch with our 'inner Flintstone' if we are not to have our best laid plans for taking charge of history thwarted.

This idea of human nature and human culture pulling inexorably in different directions is of course hardly a new one, although arguably we can give it more of a scientific base than in the past. It was anticipated by Hobbes, by Rousseau, and of course by Freud. Denis Diderot put it strikingly in his *Supplement to Bougainville's Voyage*, in 1784:

Would you like to know the condensed history of almost all our miseries? Here it is. There existed a natural man; an artificial man was introduced within this man; and within this cavern a civil war breaks out which lasts for life.[2]

It's not new . . . and I'm beginning to realize that in large measure *it's not true*. Indeed the very opposite is true. If we take a fresh look at how natural and artificial man coexist in contemporary societies, what we see in many areas is evidence not of an ongoing civil war but of a remarkable collaboration between these supposed old enemies. We see that Stone Age nature and space age culture—or at any rate modern industrial culture—very often get along just fine.

So much so, that perhaps it's time we shifted our concerns, and started emphasizing not how human nature resists desirable cultural developments but how it eases the path of some of the least desirable. For the fact is—the worry is, the embarrassment is—that human nature may sometimes be not just a collaborator, but an active *collaborateur*, with the invader.

True, once we do focus on collaboration, for the most part we'll find the result productive and benign. Think of the major achievements of modern civilization—in science, the arts, politics, communication, commerce. Every one has depended on cultural forces working with the natural gifts of human beings. Our incomparable mental gifts: the capacity for language, reasoning, socializing, understanding, feeling with others. Equally our incomparable bodily gifts: our nimble fingers, expressive faces, graceful limbs. We have only to look at our nearest relatives, the ungainly lumbering chimpanzees, whose minds can scarcely frame a concept and whose hands can hardly hold a pencil, to see how near a thing it was. Tamper with the human genome ever so little, and the glories of civilization would topple like a pack of cards.

Would you like to know the condensed history of almost all our *satisfactions*? Here it is. There existed a natural man; an artificial man was introduced within this man; and within this cavern a *creative alliance is formed* which lasts for life.

That's the good news. But what about the bad? Think of the worst achievements of civilization: war, genocide, capitalist greed, religious bigotry, alienation, drug addiction, pornography, the despoliation of the earth, the dreariness of urban life. Every one of these, too, has come about through cultural forces working with the natural gifts of individual human beings. Our incomparable talents for aggression, competition, deceit, xenophobia, superstition, obedience, love of status, demonic ingenuity. We have only to look at chimpanzees to see how nearly we could have avoided it. Chimpanzees are hardly capable even of telling a simple lie.

Would you like to know the condensed history of almost all our miseries [again]? Here it is. There existed a natural man; an artificial man was introduced within this man; and within this cavern a *perfidious conspiracy arises* which lasts for life.

Yet, you may want to object that this way of putting things makes no sense scientifically. What justification can there be for talking, even at the level of metaphor, of natural and artificial man being 'at war', or 'collaborating', or 'forming alliances'—as if human nature and human culture really do have independent goals? Don't theorists these days think in terms of gene–culture co-evolution: with human beings being by nature creatures that make culture which in turn feeds back to nature and so on, in a cycle of dependent interaction—for which the bottom line always remains natural selection?

Indeed, isn't much of human culture, including many aspects of civilization, best thought of as part of the extended human phenotype, indirectly constructed by the human genes that it helps preserve—rather as, say, a beaver's dam is indirectly constructed by the beaver's genes? No one talks about an 'alliance' between a beaver and its dam (let alone a war).

Yes, except that I need hardly say that any such view of human culture—as on a par with the beaver's dam or even the tool-making traditions of chimpanzees—ignores precisely what it is that makes human culture special. I say 'precisely what it is', and no doubt everyone would want to be precise in different ways. But what *I* mean here is that once a culture becomes the creation of thousands upon thousands of semi-autonomous individual agents, each with the capacity to act both as an originator of ideas and as a vehicle and propagandist for them, it becomes *a complex dynamical system*: a system which now has *its own developmental drive*—no longer ruled by natural selection, but ruled by whatever principles do rule such complex systems.

We are only just beginning to understand what these principles are. Nobody believes any longer that Hegelian laws dictate the course of human history. Few believe in the Marxist version (though perhaps more should—Engels's remarkable book *Dialectics of Nature* anticipated in several ways modern ideas about complexity and chaos). Even Adam Smith's hidden hand of capitalism has been relegated to the nursery. But that there *are* laws of development, theorists are increasingly confident. And certain central concepts are becoming established. Probably the most important, simple, and easy to understand is the idea of 'attractors'.

Attractors are just that—basins of attraction, places where

complex systems tend to end up *as if they were drawn there*. They are semi-stable states of equilibrium in a sea of instability—like whirlpools in the wide ocean tending to catch within them anything that strays into their fields of influence (although they do not, of course, literally pull). Weather systems, economies, fashions, ecosystems, dripping taps, traffic flow along motorways—all tend to settle into discrete attractor states.

Now, the existence of attractors does not, of course, mean that, strictly speaking, the system that exhibits these attractors has its own *goals* (certainly not consciously formulated). An economy headed for recession doesn't exactly 'want' to get there. Nonetheless, it will seem to anyone caught up in such a recession, or it might be in an artistic movement, a religious crusade, or a slide towards war, that he is part of something larger than himself with its own life force. While being one of the players whose joint effects produce the movement, he does not have to play any intentional part in it. Indeed, even in fighting it he may ironically promote it, or in embracing it oppose it.

In short, nature and culture *can* be thought of as having independent tendencies and independent energies. Diderot's natural man, *as an individual*, will go his own way; but there will be emergent forces *at the group level* creating the movement towards the cultural attractor where Diderot's artificial man is at home.

So, now's the point: whether natural man intends it or not, his going his own way is likely to have interesting and possibly quite unpredicted consequences at the cultural level. And this is because—as complexity theory makes clear—the determination of which particular states are in fact attractors, and how easily transitions occur between them, is often highly sensitive to small changes in the properties of the elements that comprise the system. Adjust the stickiness of sand particles in the desert ever so slightly, and whole dune systems will come into being or vanish. Adjust the saving or spending habits of people in the market by a few percentage points, and the whole economy can be tipped in or out of recession. So: adjust human nature in other small ways, and maybe human society as a whole can be unexpectedly switched from Victorian values to flower power, from capitalism to socialism, from theism to atheism, and so on? (Well, no: atheism is probably never an attractor!)

There. That is my preamble. And I'm ready now to address the subject I was asked to: the question of *the attractions of dictatorship*.

As Plato noted, the political landscape contains several distinct, relatively stable forms of government. He named them as: Timarchy, Oligarchy, Democracy, Tyranny, and of course his own Ideal Republic (which was itself a form of benevolent dictatorship.) Moreover, as if foreseeing complexity theory, Plato drew attention to the critical influence of human personality in rendering these states more or less stable: because, as he said, 'Societies aren't made of sticks and stones, but of men whose individual characters, by turning the scale one way or another, determine the direction of the whole'.[3]

Now, as Plato also noted, some of these attractors are nearer neighbours than others (because they are relatively similar in character), and some have higher transition probabilities between them (because there are, as it were, downhill routes connecting them). I may surprise you—though not for long—when I say I think democracy and dictatorship are close in both respects.

Dictatorship, let's be clear, does not mean any kind of authoritarian regime. It involves, as the title of this seminar suggests, 'the seduction of the masses'. Dictatorship emerges where an individual or clique takes over power, usually at a time of crisis, on behalf of and often with the support of the people, and substitutes personal authority for the rule of law. As Carl Schmitt defined it: 'dictatorship is the subordination of law to the rule of an executive power, exceptionally and temporarily commissioned in the name of the people to undertake all necessary acts for the sake of legal order'.[4]

What makes dictatorship, in this sense, close to democracy

is quite simply that both are *popular* forms of government: people love them. Indeed, sometimes dictatorships command the people's love beyond anything that democracy can normally muster. The Nuremberg rallies, the huge outpourings of love and admiration for Mao Tse Tung, even the lingering adoration shown in some quarters for fallen dictators such as Slobodan Milosevic, testify to the way people genuinely take to their hearts those who, even while oppressing them, claim to represent them.

It means that the transition from democracy to dictatorship can feel (at least to begin with) like a continuation of democracy by other means. A government of the people for the people by the people has become a government of the people for the people by something else. But it will not necessarily feel as if power has been given up, because in a sense it has not: the something else is also there by the people's will. In the case of Hitler, for example, 'the Führer's word was supposed to be law not because it was the will of a particular individual but because it was supposed to embody the will of the German people more authentically than could any representation'.[5]

The transition therefore may not be resisted as perhaps it should be. But that is by no means the full story. For there may be a positive energies as well, tending to drive society towards the beckoning attractor of dictatorship. And they may arise, I believe, from a wholly familiar (if little understood) aspect of individual human psychology: the everyday, taken-for-granted, but insidious phenomenon of *suggestibility*—the tendency of human beings, in certain circumstances, especially when stressed, to surrender their capacity for self-determination and self-control to someone else.

Suggestibility shows up as a human personality trait in two main ways. To start with, there is the simple tendency to *conform,* to adopt the wishes and opinions *of the majority* as one's own. The power of conformity was famously illustrated in Solomon Asch's experiment where he showed how group pressure can distort a person's judgement even about such a simple matter as the length of a line. Conformity has been shown to be a powerful force in contexts ranging from shopping behaviour (so that people want to buy what lots of other people buy) to mate choice (so that, yes, people want to have sex with the person lots of other people want to have sex with). We all do it, we can't help it. On the whole it doesn't get us into trouble.

But besides this, and more obviously dangerous, is the tendency to *yield to authority*, to identify with the wishes and opinions of *a leader*. This, too, shows up, for example, in shopping behaviour (people want to buy what the top people buy) and again in mate choice (people want to have sex with the person a top person chooses to have sex with). It plays a central role in the strange, but generally innocuous, phenomenon of hypnosis. It is seen more worryingly in examples of the so-called Stockholm syndrome, where individuals who find themselves in the power of a captor or abuser come to identify with and bond with the person into whose power they have fallen. It was demonstrated most dramatically in Stanley Milgram's obedience experiments, where he showed how easily ordinary people will take orders from an assumed expert even when it involves hurting another person.

Now, though these various forms of suggestibility have long been established as facts of human nature, from the evolutionary perspective they have seemed something of a puzzle. Recently, however, theorists have gone part of the way to solving this puzzle by thinking about suggestibility in the context of the *evolution of cooperation*.

Joseph Henrich and Robert Boyd have shown, with a theoretical model, that if human beings do indeed show *both* these forms of suggestibility—that's to say, if they have a psychological bias to copy the majority (which Henrich and Boyd call 'conformist transmission'), *and* a bias to copy successful individuals (which they call 'payoff-biased transmission'), then these two factors by themselves will be sufficient to account for the spread of cooperation as a stable feature of human groups.[6]

It's not hard to see, informally, why. Imagine two groups, A and B, in the first of which everyone cooperates, and in the second they do not. Assume that individuals in group A generally do better for themselves than those in group B because cooperation leads to more food, better health, and so on. Then, when individuals in group B have occasion to observe the relative success of those in group A, payoff-biased transmission will see to it that they tend to adopt these cooperative individuals as role models and begin cooperating themselves. And then, once enough individuals in group B are doing this, conformist transmission will take over and sweep the whole group internally towards cooperation.

To put this still more simply: if people follow the winners, and people follow the crowd, then they will end up following the crowd following the winners. And, hey presto, everyone's a winner!

Since being one of the winners does bring obvious benefits to individuals and their genes, it seems clear from this model why both these forms of suggestibility should in fact have been bred into human nature by natural selection. Suggestibility, far from being a design flaw in itself, will have been one of the basic adaptive traits that has underwritten the development of human social and economic life. But if this is indeed our Stone Age heritage, how does it play now? If not necessarily in the space age, then in the age of nation states and mass politics?

I would say that this is the very package of traits that, as Plato put it, 'by turning the scale one way or another, determines the direction of the whole', and continually threatens to make democracy unstable and dictatorship a particularly likely end-point.

The fact is that democracy, for which one of the essential conditions is the exercise of individual choice, is dreadfully vulnerable to the block vote. And what suggestibility will tend to do is precisely to turn independent choosers into block voters. If people follow the charismatic leader, Hitler, Mao, and people also follow the crowd, then they will end up following the crowd following the charismatic leader. And hey presto, everyone's a servant of dictatorship.

Yet what creates such charisma in the first place? In the situation that Henrich and Boyd have modelled, the thing that makes suggestibility an *adaptive* strategy is that the leaders, around whom the process crystallizes, have genuinely admirable qualities. However, it is easy to see how, once suggestibility is in place, there is the danger of a much less happy outcome, perhaps rare until the advent of modern propaganda, but always latent. This is that *a leader's charisma will derive from nothing else than the support he is seen to be getting from the crowd.* So that a figure who would otherwise never be taken seriously can become the subject of runaway idolatry. Just as a film star can become famous for little other than being famous, a leader can become popular by virtue of little other than his popularity.

Wherein lay Hitler's mesmerizing power? Maybe it was due to some of the personal qualities that Kershaw identifies—rhetorical skill, raging temper, paranoia, vanity. But maybe it was due to nothing other than the luck of being the focus of a critical mass of zealots at a critical time.

Remember that shocking newspaper headline (see Chapter 22): 'Archduke Ferdinand Still Alive: First World War a Mistake'? Perhaps we could have capped it with this: 'Second World War a Mistake: Hitler a Twerp'. A twerp. But nonetheless, to that young girl and so many others of her compatriots whose own breath created the twisting wind that sent him skywards, 'a great man, a genius, a person sent to us from heaven'.

It can happen with ideas as well. Take the case of Soviet Communism. J. M. Keynes wrote: 'How a doctrine so illogical and so dull can have exercised so powerful and enduring an influence over the minds of men, and, through them, the events of history must always remain a portent to historians of opinion.'[7] A portent to students of human nature, too.

Notes

Chapter 1. On Taking Another Look

First published in John Brockman and Katinka Matson, eds., *How Things Are: A Science Tool-Kit for the Mind* (New York: William Morrow, 1995), 177–82.

  1. Arthur Conan Doyle, *The Sign of Four* (1890), ch. 6.

  2. Richard L. Gregory, *The Intelligent Eye* (London: Weidenfeld & Nicolson, 1970), 55.

Chapter 2. One Self

Lecture for the Dutch arts organization, De Balie, Amsterdam, December 1997. Published in *Social Research*, 67 (2000), 32–9.

  1. Gottlob Frege, 'The Thought: A Logical Inquiry', in P.F. Strawson, ed., *Philosophical Logic* (Oxford: Oxford University Press, 1967), 67.

  2. Marcel Proust, *Swann's Way*, vol. i of *Remembrance of Things Past*, trans. C. K. Scott Moncrieff and T. Kilmartin (London: Chatto & Windus, 1981), 5.

Chapter 3. What is your Substance?

Broadcast on BBC2 television, autumn 1982.

  1. Walter H. Pater, 1873, *The Renaissance* (London: Macmillan, 1873), 119.

  2. Plato, *The Republic*, trans. H. D. P. Lee (Harmondsworth: Penguin, 1955), 137.

Chapter 4. Speaking for Our Selves

Originally commissioned, but not published, by the *New York Review of Books*; later published in *Raritan*, 9: 1 (1989), 68–98.

  1. For the statistics cited on MPD, see F. W. Putnam et al., 'The Clinical Phenomenology of Multiple Personality Disorder: Review of 100 Recent Cases', *Journal of Clinical Psychiatry*, 47 (1986), 285–93.

2. DSM-III is *Diagnostic and Statistical Manual III* (Washington, D.C.: American Psychiatric Association, 1980).

3. Among the autobiographical accounts, by far the best written is Sylvia Fraser, *My Father's House* (New York: Ticknor & Fields, 1988). Many short case histories have been published in the clinical literature. J. Damgaard, S. Van Benschoten, and J. Fagan, 'An Updated Bibliography of Literature Pertaining to Multiple Personality', *Psychological Reports*, 57 (1985), 131–7, provides a comprehensive bibliography. For a more sceptical treatment, see Thomas A. Fahy, 'The Diagnosis of Multiple Personality Disorder: A Critical Review', *British Journal of Psychiatry*, 153 (1988), 509–606.

4. David Hume, 'Of Miracles' (1748), in *Essays and Treatises on Several Subjects* (Edinburgh: Bell & Bradfute, 1809), ii. 115–38, at 121.

5. See Daniel Dennett, 'Why We Are All Novelists', *Times Literary Supplement*, 16–22 September 1988, 1016–29.

6. Eugene Marais, *The Soul of the White Ant* (London: Methuen, 1937). Douglas R. Hofstadter has developed the analogy between mind and ant colony in the 'Prelude . . . Ant Fugue' flanking ch. 10 of *Gödel, Escher, Bach* (New York: Basic Books, 1979). The 'distributed control' approach to designing intelligent machines has in fact had a long history in Artificial Intelligence, going back as far as Oliver Selfridge's early 'Pandemonium' model of 1959, and finding recent expression in Marvin Minsky's *The Society of Mind* (New York: Simon & Schuster, 1985).

7. The social function of self-knowledge has been stressed particularly by Nicholas Humphrey in *Consciousness Regained* (Oxford: Oxford University Press, 1983), and in *The Inner Eye* (London: Faber & Faber, 1986). For a suggestive discussion of 'active symbols' as something not unlike our notion of a Head of Mind, see Douglas R. Hofstadter, 'Prelude . . . Ant Fugue', and *Metamagical Themas* (New York: Basic Books, 1985), esp. 646–65.

8. C. S. Peirce, 'What Pragmatism Is', *Monist* (1905), reprinted as 'The Essentials of Pragmatism' in *Philosophical Writings of Peirce*, ed. J. Buchler (New York: Dover, 1955), 258.

9. Heinz Kohut, 'On Courage' (early 1970s), reprinted in H. Kohut, *Self Psychology and the Humanities*, ed. Charles B. Strozier (New York: W. W. Norton, 1985), 33.

10. Robert Jay Lifton, *The Broken Connection* (New York: Simon & Schuster, 1979); *The Nazi Doctors* (New York: Basic Books, 1986).

11. *S4OS—Speaking for Our Selves: A Newsletter by, for, and about People with Multiple Personality*, P.O. Box 4830, Long Beach, Calif. 90804, published quarterly between October 1985 and December 1987, when publication was suspended (temporarily, it was hoped) due to a personal crisis in the life of the editor. Its contents have unquestionably been the

sincere writings and drawings of MPD patients, often more convincing—
and moving—than the many more professional autobiographical accounts
that have been published.

12. Frank W. Putnam, 'The Psychophysiologic Investigation of Multiple
Personality Disorder', *Psychiatric Clinics of North America*, 7 (March
1984), 31–9; Scott D. Miller, 'The Psychophysiological Investigation of
Multiple Personality Disorder: Review and Update', in B. G. Braun, ed.,
Dissociative Disorders (Chicago: Dissociative Disorders Program, 1988),
113; Chicago; Mary Jo Nissen, James L. Ross, Daniel B. Willingham,
Thomas B. MacKenzie, and Daniel L. Schachter, 'Memory and Awareness
in a Patient with Multiple Personality Disorder', *Brain and Cognition*, 8
(1988), 117–34, and 'Evaluating Amnesia in Multiple Personality Disorder', in R. M. Klein and B. K. Doane, eds., *Psychological Concepts and
Dissociative Disorders* (Hillsdale, N.J.: Erlbaum, 1994).

13. On incipient MPD in children, see David Mann and Jean Goodwin,
'Obstacles to Recognizing Dissociative Disorders in Child and Adolescent
Males', in Braun, ed., *Dissociative Disorders*, 35; Carole Snowden, 'Where
Are All the Childhood Multiples? Identifying Incipient Multiple Personality
in Children', in Braun, ed., *Dissociative Disorders*, 36; Theresa K. Albini,
'The Unfolding of the Psychotherapeutic Process in a Four Year Old Patient
with Incipient Multiple Personality Disorder', in Braun, ed., *Dissociative
Disorders*, 37.

14. For a fascinating discussion of how individuals may mould themselves to fit 'fashionable' categories, see Ian Hacking, 'Making Up People',
in Thomas C. Heller, Morton Sosna, and David E. Wellbery, eds.,
Reconstructing Individualism (Stanford, Calif.: Stanford University Press,
1986), 222–36.

15. W. S. Taylor and M. F. Martin, 'Multiple personality', *Journal of
Abnormal and Social Psychology*, 39 (1944), 281–300.

Chapter 5. Love Knots

First published in the *Guardian*, 11 February 1987.

1. Donald Attwater, *The Penguin Dictionary of Saints* (Harmondsworth: Penguin, 1965).

2. *Encyclopaedia Britannica*, 14th edn. (Chicago: Encyclopaedia
Britannica, Inc., 1929). The subsequent extracts come from the same edition.

3. John Donne, 'An Epithalamion, Or Marriage Song on the Lady
Elizabeth and Count Palatine being Married on St Valentine's Day', *Poetical
Works*, ed. Sir Herbert Grierson (Oxford: Oxford University Press, 1933),
113.

Chapter 6.   Varieties of Altruism

First published in *Social Research*, 64 (1997), 199–209.

1. W. D. Hamilton, 'The Evolution of Altruistic Behaviour', *American Naturalist*, 97 (1963), 354–6.
2. Robert L. Trivers, 'The Evolution of Reciprocal Altruism', *Quarterly Review of Biology*, 46 (1971), 35–57.
3. W. D. Hamilton, *Narrow Roads of Gene Land: The Collected Papers of W. D. Hamilton* (Oxford: W. H. Freeman, 1996), i. 263.
4. Trivers, 'The Evolution of Reciprocal Altruism'.
5. Steven I. Rothstein, 1980, 'Reciprocal Altruism and Kin Selection Are Not Clearly Separable Phenomena', *Journal of Theoretical Biology*, 87 (1980), 255–61.
6. Richard Dawkins, *The Selfish Gene* (Oxford: Oxford University Press, 1976), 96; also *The Extended Phenotype* (Oxford: Oxford University Press, 1982), 155.
7. Frans de Waal, *Good Natured: The Origins of Right and Wrong in Humans and Other Animals* (Cambridge, Mass.: Harvard University Press, 1996).
8. Charles Kingsley, *The Water-Babies* (1863), ch. 5.

Chapter 7.   The Uses of Consciousness

An abbreviated version of the James Arthur Memorial Lecture, American Museum of Natural History, New York, 1987. Parts are taken from Nicholas Humphrey, *The Inner Eye* (London: Faber & Faber, 1986).

1. Denis Diderot, *Elements of Physiology* (1774–80), in *Diderot: Interpreter of Nature*, trans. and ed. Jonathan Kemp (London: Lawrence & Wishart, 1937), 136.
2. Ibid. 139.
3. Charles Darwin, 'B Notebook' (1838), B232, in Paul H. Barrett, ed., *Metaphysics, Materialism and the Evolution of Mind* (Chicago: University of Chicago Press, 1980), 187.
4. Diderot, *Elements of Physiology*, 136.
5. J. B. Watson, *Behaviorism* (London: Routledge & Kegan Paul, 1928), 6.
6. Ludwig Wittgenstein, *Philosophical Investigations*, trans. G. E. M. Anscombe (Oxford: Blackwell, 1958).
7. Diderot, *Elements of Physiology*, 139.
8. L. Weiskrantz, *Blindsight* (Oxford: Oxford University Press, 1986).
9. Diderot, *Elements of Physiology*, 136.
10. Ibid. 134.
11. Ludwig Wittgenstein, *Tractatus Logico-Philosophicus* (London: Kegan Paul, 1922), 189.

12. Denis Diderot, *Promenade of a Skeptic* (1747), in *Diderot: Interpreter of Nature*, trans. and ed. Jonathan Kemp (London: Lawrence & Wishart, 1937), 28.

13. See the brilliant discussion of self-reference in Douglas R. Hofstadter, *Gödel, Escher, Bach* (New York: Basic Books, 1979).

14. Donald R. Griffin, *Listening in the Dark* (New Haven, Conn.: Yale University Press, 1958).

15. Cited by A. V. Hill, *The Ethical Dilemma of Science* (New York: Rockefeller Institute, 1960), 135.

16. Wittgenstein, *Philosophical Investigations*, Pt. I, § 294.

17. Diderot, *Elements of Physiology*, 138.

18. David Premack and Ann Premack, *The Mind of an Ape* (New York: W. W. Norton, 1983).

Chapter 8. Farewell, Thou Art Too Dear

Broadcast on BBC2 television, autumn 1982.

1. W. H. Auden, 'Writing', in *The Dyer's Hand and Other Essays* (London: Faber & Faber, 1963).

Chapter 9. How to Solve the Mind–Body Problem

First published in *Journal of Consciousness Studies*, 7 (2000), 5–20. Commentaries and my reply appear in the same issue of the journal.

1. Denis Diderot, *On the Interpretation of Nature* (1754), XXX, in *The Irresistible Diderot*, ed. J. H. Mason (London: Quartet, 1982), 66.

2. Ibid. XLV, 68.

3. David J. Chalmers, *The Conscious Mind* (Oxford: Oxford University Press, 1996).

4. Colin McGinn, 'Can We Solve the Mind–Body Problem?', *Mind*, 98 (1989), 349–66.

5. J. W. von Goethe, *Conversations with Eckermann*, 11 April 1827, trans. John Oxenford (Bohn's Standard Library, 1850).

6. Diderot, *On the Interpretation of Nature*, XXIII, 46.

7. Compare the discussion of the same problem by J. Scott Kelso, *Dynamical Patterns: The Self-Organisation of Brain and Behavior* (Cambridge, Mass.: MIT Press, 1995), 29.

8. A. S. Ramsey, *Dynamics* (Cambridge: Cambridge University Press, 1954), 42.

9. Peter Munz, 'The Evolution of Consciousness: Silent Neurons and the Eloquent Mind', *Journal of Social and Evolutionary Systems*, 20 (1997), 7–28.

10. Colin McGinn, 'Consciousness and Cosmology: Hyperdualism

Ventilated', in M. Davies and G. W. Humphreys, eds., *Consciousness* (Oxford: Blackwell, 1993), 155–77, at 160.

11. Daniel C. Dennett, *Consciousness Explained* (New York: Little Brown, 1991).

12. Roger Penrose, *The Emperor's New Mind* (Oxford: Oxford University Press, 1989).

13. Isaac Newton, 'A Letter from Mr. Isaac Newton . . . Containing his New Theory about Light and Colours', *Philosophical Transactions of the Royal Society*, 80 (1671), 3075–87, at 3085.

14. Daniel C. Dennett, 'Quining Qualia', in A. J. Marcel and E. Bisiach, eds., *Consciousness in Contemporary Science* (Oxford: Clarendon Press, 1988), 42–77.

15. F. Crick and C. Koch, 'The Unconscious Homunculus', *Neuro-Psychoanalysis*, 2 (2000), 1–10.

16. Susan Greenfield, 'How Might the Brain GenerateConsciousness?', in Steven. Rose, ed., *From Brains to Consciousness* (London: Allen Lane, 1998), 210–27, at 210.

17. Antonio Damasio, *The Feeling of What Happens* (London: Heinemann, 2000), 9.

18. Thomas Reid, *Essays on the Intellectual Powers of Man* (1785), ed. D. Stewart (Charlestown: Samuel Etheridge, 1813), Pt. II, ch. 17, p. 265, and ch. 16, p. 249.

19. Nicholas Humphrey, *A History of the Mind* (London: Chatto & Windus, 1992).

20. Reid, *Essays on the Intellectual Powers of Man*, Pt. II, ch. 17, p. 265, and ch. 16, p. 249.

21. Ned Block, 'On a Confusion about a Function of Consciousness', *Behavioral and Brain Sciences*, 18 (1995), 227–47; John R. Searle, *The Rediscovery of the Mind* (Cambridge, Mass.: MIT Press, 1992); see also my discussion in Nicholas Humphrey, 'Now You See It, Now You Don't', *Neuro-Psychoanalysis*, 2 (2000), 14–17.

22. V. S. Ramachandran and W. Hirstein, 'Three Laws of Qualia: What Neurology Tells Us about the Biological Functions of Consciousness', *Journal of Consciousness Studies*, 4 (1997), 429–57.

23. Thomas Reid, *An Inquiry into the Human Mind* (1764), ed. D. Stewart (Charlestown: Samuel Etheridge, 1813), 112.

24. Humphrey, *A History of the Mind*.

25. Nicholas Humphrey, 'The Thick Moment', in J. Brockman, ed., *The Third Culture* (New York: Simon & Schuster, 1995), 198–208.

26. Denis Diderot, *Elements of Physiology* (1774–80), in *Diderot: Interpreter of Nature*, trans. and ed. Jonathan Kemp (London: Lawrence & Wishart, 1937), 139.

Chapter 10. The Privatization of Sensation

An abbreviated version of my keynote lecture at the Third Tucson Conference on Consciousness, 1998, published in S. R. Hameroff, A. W. Kaszniak, and D. J. Chalmers, eds., *Toward a Science of Consciousness III* (Cambridge, Mass.: MIT Press, 1999), 247–58; also in C. Heyes and L. Huber, eds., *The Evolution of Cognition* (Cambridge, Mass.: MIT Press, 2000), 241–52.

1. William Empson, *Seven Types of Ambiguity*, 3rd edn. (Harmondsworth: Penguin, 1961), 28.

2. Thomas Reid, *Essays on the Intellectual Powers of Man* (1785), ed. D. Stewart (Charlestown: Samuel Etheridge, 1813), 265.

3. Nicholas Humphrey, *A History of the Mind* (London: Chatto & Windus, 1992); Anthony J. Marcel, 'Phenomenal Experience and Functionalism', in A. J. Marcel and E. Bisiach, eds., *Consciousness in Contemporary Science* (Oxford: Clarendon Press, 1988), 121–58; Richard L. Gregory, 'What Do Qualia Do?', *Perception*, 25 (1996), 377–8.

4. Humphrey, *A History of the Mind*, 73.

5. John Locke, *An Essay Concerning Human Understanding* (1690), ed. P. H. Nidditch (Oxford: Clarendon Press, 1975), Pt. II, ch. 32.

6. Ludwig Wittgenstein, *Philosophical Investigations*, trans. G. E. M. Anscombe (Oxford: Blackwell, 1958), Pt. I, § 272.

7. Daniel C. Dennett, 'Quining Qualia', in A. J. Marcel and E. Bisiach, eds., *Consciousness in Contemporary Science* (Oxford: Clarendon Press, 1988), 42–77; however, compare Dennett's more nuanced position in *Consciousness Explained* (New York: Little Brown, 1991).

8. Philip Steadman, *The Evolution of Designs* (Cambridge: Cambridge University Press, 1979).

9. Humphrey, *A History of the Mind*; Nicholas Humphrey, 'The Thick Moment', in J. Brockman, ed., *The Third Culture* (New York: Simon & Schuster, 1995), 198–208.

10. Euan Macphail, 'The Search for a Mental Rubicon', in C. Heyes and L. Huber, eds., *The Evolution of Cognition* (Cambridge, Mass.: MIT Press, 2000).

11. Gottlob Frege, 'The Thought: A Logical Inquiry', in P. F. Strawson, ed., *Philosophical Logic* (Oxford: Oxford University Press, 1967), 27.

12. Rudyard Kipling, 'A Smuggler's Song' (1914).

Chapter 11. Mind in Nature

First published in the *Guardian*, 10 December 1986.

1. William James, *The Principles of Psychology* (New York: Henry Holt, 1890), i. 8.

2. Jonathan Schull, 'Are Species Intelligent?', *Behavioral and Brain Sciences*, 3 (1990), 417–58.

3. *Antony and Cleopatra*, V. ii. 88–9.

Chapter 12. *Cave Art, Autism, and Evolution*

First published, with commentaries and a reply, in *Cambridge Archaeological Journal*, 8 (1998), 165–91; reprinted in *Journal of Consciousness Studies*, 6 (1999), 116–43. Nadia's drawings are reproduced with kind permission of Lorna Selfe and Academic Press.

1. E. H. Gombrich, 'The Miracle at Chauvet', *New York Review of Books*, 14 November 1996, 8–12, at 8.

2. Steven J. Mithen, quoted by Tara Patel, 'Stone Age Picassos', *New Scientist*, 13 July 1996, 33–5, at 33.

3. Erich Neumann, *Art and the Creative Unconscious*, trans. Ralph Manheim (Princeton, N.J.: Princeton University Press, 1971), 86.

4. Terrence W. Deacon, *The Symbolic Species: The Co-Evolution of Language and the Brain* (New York: Norton, 1997), 374.

5. Steven J. Mithen, 'Looking and Learning: Upper Palaeolithic Art and Information Gathering', *World Archaeology*, 19 (1988), 297–327, at 314.

6. Gombrich, 'The Miracle at Chauvet', 10.

7. Lorna Selfe, *Nadia: A Case of Extraordinary Drawing Ability in an Autistic Child* (London: Academic Press, 1977; Lorna Selfe, *Normal and Anomalous Representational Drawing Ability in Children* (London: Academic Press, 1983); Lorna Selfe, 'Anomalous Drawing Development: Some Clinical Studies', in N. H. Freeman and M. V. Cox, eds., *Visual Order: The Nature and Development of Pictorial Representation* (Cambridge: Cambridge University Press, 1985), 135–54.

8. Selfe, *Nadia*, 127.

9. Jean Clottes, 'Epilogue', in Jean-Marie Chauvet, Eliette Brunel Deschamps, and Christian Hilaire, *Dawn of Art: The Chauvet Cave* (New York: Harry N. Abrams, Inc., 1996), 114.

10. Ibid.

11. See Uta Frith and Francesca Happé, 'Autism: Beyond "Theory of Mind"', *Cognition*, 50 (1994), 115–32.

12. Amitta Shah and Uta Frith, 'An Islet of Ability in Autistic Children: A Research Note', *Journal of Child Psychology and Psychiatry*, 24 (1983), 611–20.

13. Frith and Happé, 'Autism'; see also L. Pring, B. Hermelin, and L. Heavey, 'Savants, Segments, Art and Autism', *Journal of Child Psychology and Psychiatry*, 36 (1995), 1065–76.

14. Selfe, *Nadia*, note to Pl. 33.

15. Ian Tattersall, *Becoming Human: Evolution and Human Uniqueness* (New York: Harcourt Brace, 1998), 16.

16. Selfe, 'Anomalous Drawing Development', 140.

17. Selfe, *Normal and Anomalous Representational Drawing Ability in Children*, 201.

18. Robin Dunbar, *Grooming, Gossip and the Evolution of Language* (Cambridge, Mass.: Harvard University Press, 1996).

19. Steven J. Mithen, *The Prehistory of the Mind* (London: Thames & Hudson, 1996).

20. Ibid. 163.

21. J. M. Keynes, 'Newton the Man', in *Newton Tercentenary Celebrations* (Cambridge: Cambridge University Press, 1947), 34–7.

22. Nicholas Humphrey, 'Species and Individuals in the Perceptual World of Monkeys', *Perception*, 3 (1974), 105–14.

23. Elizabeth Newson, 'Postscript' to Selfe, *Nadia*, 129.

24. Jean Clottes, 'Thematic Changes in Upper Palaeolithic Art: A View from the Grotte Chauvet', *Antiquity*, 70 (1996), 276–88.

25. Paul G. Bahn, Paul Bloom and Uta Frith, Ezra Zubrow, Steven Mithen, Ian Tattersall, Chris Knight, Chris McManus, and Daniel C. Dennett, *Cambridge Archaeological Journal*, 8 (1998), 176–91.

26. Robert Darnton, *The Great Cat Massacre* (Harmondsworth: Penguin, 1985), 12,

27. Selfe, *Nadia*, 8 and 103.

28. Stephen Wiltshire's case is described in Oliver Sacks, 'Prodigies', in *An Anthropologist on Mars* (London: Picador, 1995), 179–232.

29. Daniel C. Dennett, *Consciousness Explained* (New York: Little Brown, 1992); Daniel C. Dennett, *Kinds of Minds* (New York: Basic Books, 1996).

30. Allen W. Snyder and Mandy Thomas, 'Autistic Artists Give Clues to Cognition', *Perception*, 26 (1997), 93–6, at 95.

31. See Sacks, 'Prodigies'.

32. E. H. Gombrich, *Art and Illusion* (London: Phaidon, 1960), 18.

33. Selfe, *Nadia*, 8.

34. Geoffrey Miller, 'How Mate Choice Shaped Human Nature', in Charles Crawford and Denis Krebs, eds., *Handbook of Evolutionary Psychology* (Mahwah, N.J.: Erlbaum, 1998).

35. Steven J. Mithen, report in *New Scientist*, 9 May 1998, 16.

Chapter 13. Scientific Shakespeare

First published in the *Guardian*, 26 August 1987.

1. C. P. Snow, *The Two Cultures and the Scientific Revolution* (Cambridge: Cambridge University Press, 1959).

2. Thomas Maude, *Wensley-Dale . . . A Poem* (1772), stanza 23, note; probably apocryphal.

Chapter 14. *The Deformed Transformed*

Based on a radio talk, 'The Mother of Invention', originally broadcast on BBC Radio 3, November 1979.

1. W. Derham, *Physico-Theology: Or a Demonstration of the Being and Attributes of God from His Works of Creation*, 6th edn. (London, 1723), 321.

2. W. Ikle, D. N. Ikle, S. G. Morland, L. M. Fanshaw, N. Waas, and A. M. Rosenberg, 'Survivors of Neonatal Extracorporeal Membrane Oxygenation at School Age: Unusual Findings on Intelligence Testing', *Developmental Medical Child Neurology*, 41 (1999), 307–10.

3. Narinder Kapur, 'Paradoxical Functional Facilitation in Brain–Behaviour Research: A Critical Review', *Brain*, 119 (1996), 1775–90.

4. Geoffrey Miller, 'Sexual Selection for Indicators of Intelligence', in *The Nature of Intelligence* (Novartis Foundation Symposium 233) (Chichester: Wiley, 2000).

5. S. W. Gangestad and J. A. Simpson, 'The Evolution of Mating: Trade-Offs and Strategic Pluralism', *Behavioural and Brain Sciences*, 23 (2000).

6. Charles Darwin, *On the Origin of Species*, 2nd edn. (London: Murray, 1861), 149.

7. Derham, *Physico-Theology*, 268.

8. Editorial on 'Irritant Woods', *British Medical Journal*, 16 October 1976, 900.

9. Nicholas Humphrey, 'The Social Function of Intellect', in P. P. G. Bateson and R. A. Hinde, eds., *Growing Points in Ethology* (Cambridge: Cambridge University Press, 1976), 303–17, at 311.

10. Nehemiah Grew, *Cosmologica Sacra* (1701), Bk. 1, ch. 5, sect. 25, quoted by Derham, *Physico-Theology*, 292.

11. William Shakespeare, *As You Like It*, II. i. 12–13.

12. Friedrich Nietzsche, *The Gay Science* (1882), in *A Nietzsche Reader,* ed. R. J. Hollingdale (Harmondsworth: Penguin, 1977), 100.

13. Martin Amis, *Night Train* (New York: Harmony Books, 1998).

14. See Denis Brian, *Einstein: A Life* (London: Wiley, 1996), 61.

15. In the simplest case, as I've outlined it, we will be looking for evidence of the original genetic traits being replaced by invented and culturally transmitted ones. However, in reality this picture may have become blurred somewhat in the longer course of human evolution. The reason is that there is a well established rule in evolution, to the effect that when the same kind of learning occurs generation after generation, invented or learned traits tend over time to get 'assimilated' into the genome, so that eventually they

themselves become genetically based (the 'Baldwin effect', see Chapter 11). We must be prepared, therefore, for the possibility that a genetic trait has been replaced by a learned one as a result of the Grew effect, but that this new trait may nonetheless itself today be largely genetic.

16. Desmond Morris, *The Naked Ape* (London: Cape, 1967).

17. Anyone who has watched pigmy chimpanzees in sex play, or for that matter anyone who has ever fondled a pet cat, will realize that tactile stimulation can be pleasurable enough even through a hairy pelt. And anyone who has observed cheetahs or lions hunting on the savannah or gazelles outrunning them, will realize that it is possible for hair-covered animals to keep up a sustained chase without suffering major problems of overheating. But, in any case, it is not even clear that the net result of hairlessness for ancestral humans would have been to reduce the danger of overheating, since one of the prices that has to be paid for hairlessness is a black-pigmented skin, to prevent damage to the body's biochemistry from the ultraviolet light that now falls directly on the body surface: and black skins of course absorb more infra-red radiation and so tend to heat up faster in sunlight (besides cooling faster at night).

18. *Cambridge Encyclopedia of Human Evolution,* ed. Steve Jones, Robert Martin, and David Pilbeam (Cambridge: Cambridge University Press, 1992), 49.

19. Ian Tattersall, *Becoming Human* (New York: Harcourt Brace, 1998), 148.

20. D. N. Farrer, 'Picture Memory in the Chimpanzee', *Perceptual and Motor Skills,* 25 (1967), 305–15.

21. New evidence that chimpanzees have picture memory is contained in Nobuyuki Kawai and Tetsuro Matsuzawa, 'Numerical Memory Span in a Chimpanzee', *Nature,* 403 (2000), 39–40.

22. Daniel Povinelli, *Folk Physics for Apes* (Oxford: Oxford University Press, 2000) 308–11.

23. A. R. Luria, *The Mind of a Mnemonist,* trans. Lynn Solotaroff (London: Cape, 1969), 59.

24. Jorge Luis Borges invented perhaps an even more startling example in his story 'Funes the Memorious' (*Ficciones,* 1956). 'He knew by heart the forms of the southern clouds at dawn on 30 April 1882, and could compare them in his memory with the mottled streaks on a book in Spanish binding he had seen only once, and with the outlines of the foam raised by an oar in the Río Negro the night before the Quebracho uprising. These memories were not simple ones; each visual image was linked to muscular sensations, thermal sensations, etc. He could reconstruct all his dreams, all his half-dreams. Two or three times he had reconstructed a whole day; he never hesitated, but each reconstruction had required another whole day. He told me: "I alone have more memories than all mankind has probably had since the

world has been the world." . . . I suspect, however, that he was not very capable of thought. To think is to forget differences, generalise, make abstractions. In the teeming world of Funes, there were only details, almost immediate in their presence.' ('Funes the Memorious', in Jorge Luis Borges, *Labyrinths*, ed. Donald A. Yates and James E. Irby (Harmondsworth: Penguin, 1970), 92.)

25. Jonathan W. Schooler and Tonya Y. Engstler-Schooler, 'Verbal Overshadowing of Visual Memories: Some Things Are Better Left Unsaid', *Cognitive Psychology*, 22 (1990), 36–71.

26. This is the field of 'cognitive archaeology'. The sort of thing that can be done, for instance, is to follow the design of stone tools, and look for evidence of when their makers first begin to think of each tool as being of a definite *kind*—a 'hammer', a 'chopper', a 'blade' (compare our own conceptual expectations of 'knife', 'fork', and 'spoon'). See Steven J. Mithen, *The Prehistory of the Mind* (New York: Thames & Hudson, 1996).

27. This seems too recent? It does seem surprisingly recent. Not much time for a genetic trait to spread through the entire human population. I think the best way to understand it is, in fact, to attribute at least part of the change to non-genetic means—to the snowballing of a meme; see also the discussion in Chapter 12.

28. It is for just this reason that modern-day computer programmers, when updating a complex program, generally prefer to suppress obsolete bits of the old program rather than excise them—with the interesting result that the latest version of the program still contains large swathes of earlier, obsolete versions in a silenced form. The software for WordPerfect 9.0, for instance, almost certainly has long strings of the programs for WordPerfect 1.0 to 8.0 hidden in its folders. We do not know whether the DNA of *Homo sapiens sapiens* still contains long strings of the DNA of earlier human versions hidden in its folders (perhaps as what is called junk DNA)—*Homo erectus*, *Homo habilis*, *Australopithecus*, etc. But, since what makes sense for modern-day programmers of software has almost certainly always made sense for natural selection as a programmer of DNA, we should not be surprised to find that this *is* so.

29. C. Holden, 'Probing Nature's Hairy Secrets', *Science*, 268 (1995), 1439.

30. B. L. Miller, J. Cummings, F. Mishkin, K. Boone, F. Prince, M. Ponton, and C. Cotman, 'Emergence of Artistic Talent in Frontotemporal Dementia', *Neurology*, 51 (1998), 978–82.

31. Eugene Marais, *The Soul of the Ape* (New York: Atheneum, 1969).

32. J. Klepkamp, A. Riedel, C. Harper, and H.-J. Ktretschman, 'Morphometric Study on the Postnatal Growth of the Visual Cortex of Australian Aborigines and Caucasians', *Journal of Brain Research*, 35 (1994), 541–8.

33. There would seem to be two possibilities. One is the one I have been pushing, namely that in normal brains there is active inhibition of memory, which has been put there as a designed-in feature in order to limit memory. In this case, if there are particular individuals in whom the influence is lifted, this will likely be because the specific genes responsible for the inhibition are not working, or the pathways by which these genes work are damaged.

But the alternative possibility, which I have also allowed for above, is that in normal brains there is competition for resources between memory and other mental functions such as language, so that, far from being a designed-in feature, the limits on memory are merely a side effect of the brain trying to do several jobs at the same time. In this case, if the influence is lifted, this will more likely be because the competition from the other mental operations is reduced or eliminated.

Which is it? Most theorists, apart from me, would favour the competition explanation. I do not say they are wrong. It is certainly true that, with the artistic patients with dementia, their skills emerge only when their language falls away. Autistic savant children almost always have retarded language; and if their language begins to improve, their abilities for drawing or remembering usually decline (as in fact happened to Nadia). Even in normal children who have eidetic imagery, this tends to disappear around the same time as language is developing. All this does suggest that heightened memory in such cases is due to the lack of competition from language—and that the diminishment of language has come first.

Several researchers have in fact argued that the 'islands of exceptional ability' in autism are a direct result of brain resources being freed up because they are not being used in the ways they would be normally. For arguments to this effect, see the review by Rita Carter, 'Tune In, Turn Off', *New Scientist*, 9 October 1999, 30–4.

All the same, I am not sure. For a start, some of the evidence just cited could equally well be given the reverse interpretation, namely that the lack of language is due to an over-capacious memory—and that it is actually the heightening of memory that comes first. But the best evidence that, in some cases anyway, it all begins, rather than ends, with heightened memory comes from those rare cases such as S. in whom the release of memory occurs without any accompanying defects in language.

34. Luria, *The Mind of a Mnemonist*, 63.

35. Geoffrey Miller, *The Mating Mind* (London: Weidenfeld & Nicolson, 2000).

36. Ibid.

37. See, for example, the discussion of mistakes by Daniel Dennett, 'How to Make Mistakes', in John Brockman and Katinka Matson, eds., *How Things Are* (London: Weidenfeld & Nicolson, 1995), 137–44.

38. CNN radio news report, New York, 1 August 1998.
39. Matthew 19: 30.
40. Lord Byron, *The Deformed Transformed* (1824), I. i. 313–21.

Chapter 15. *Tall Stories from Little Acorns*

First published in the *Guardian*, 1 July 1987.
1. F. A. Keynes, *Gathering Up the Threads* (Cambridge: W. Heffer & Sons, 1950)
2. W. H. Auden, 'Writing', in *The Dyer's Hand and Other Essays* (London: Faber & Faber, 1963).

Chapter 16. *Behold the Man*

Inaugural lecture, New School for Social Research, New York, November 1995; partly based on chs. 13 and 14 of Nicholas Humphrey, *Soul Searching: Human Nature and Supernatural Belief* (London: Chatto & Windus, 1995), published in the USA as *Leaps of Faith: Science, Miracles, and the Search for Supernatural Consolation* (New York: Basic Books, 1996).
1. Jerry Fodor, *Representations* (Cambridge, Mass.: MIT Press, 1981), 316.
2. Bertrand Russell, *Sceptical Essays* (London: Allen & Unwin, 1928), opening paragraph.
3. Advertisement for the Perrott–Warrick Fellowship, Darwin College, Cambridge, 1991.
4. Biblical accounts of Jesus's life cannot of course be trusted to be historically accurate: they were written many years after the supposed events, and in any case they were the work of men who had an obvious interest in embellishing their stories. We cannot expect to be able to subject the reports of Jesus's miracles to the kind of cross-examination we might want to give a more reliable history. Nevertheless, it is a valuable exercise to consider what we shold make of this testamentary evidence *if* we had reason to take it at face value.
5. See accounts in Morton Smith, *Jesus the Magician* (London: Gollancz, 1978); Paul Kurtz, *The Transcendental Temptation* (Buffalo, N.Y.: Prometheus Books, 1986).
6. See Smith, *Jesus the Magician*, 90.
7. Ibid. 92–3.
8. Origen, *Against Celsus*, quoted ibid. 83.
9. Luke 4: 23.
10. Mark 6: 5.
11. Smith, *Jesus the Magician*, 141.

12. Matthew 13: 58.

13. John 4: 48.

14. Matthew 27: 40, 42.

15. Ernst Becker, *The Denial of Death* (New York: Free Press, 1973), 18.

16. Romans 1: 3.

17. Isaiah 7: 14.

18. Micah 5: 2.

19. Numbers 24: 17.

20. Robin Lane Fox, *The Unauthorized Version* (London: Viking, 1991); A. N. Wilson, *Jesus* (London: Sinclair-Stevenson, 1992); also sources cited in Smith, *Jesus the Magician*, and Kurtz, *The Transcendental Temptation*.

21. J. Mayo, O. White, and H. J. Eysenck, 'An Empirical Study of the Relation between Astrological Factors and Personality', *Journal of Social Psychology*, 105 (1978), 229–36.

22. H. J. Eysenck and D. K. B. Nias, *Astrology: Science or Superstition?* (New York: St Martin's Press, 1982).

23. Wilson, *Jesus*, 73–86.

24. Luke 2: 48, 49.

25. Nicholas Humphrey, 'Folie à Deux', *Guardian*, 8 April 1987. (The boy's name has been changed to preserve anonymity.)

26. Uri Geller, quoted by David Marks and Richard Kammann, *The Psychology of the Psychic* (Buffalo, N. Y.: Prometheus Books, 1980), 90; see also Uri Geller, *My Story* (New York: Praeger, 1975).

27. Arthur Koestler, *Arrow in the Blue: An Autobiography* (New York: Macmillan, 1952), 36.

28. Arthur Koestler, quoted by Adam Smith in *New York* magazine, 27 December 1976.

29. Uri Geller, quoted by Merrily Harpur, 'Uri Geller and the Warp Factor', *Fortean Times*, 78 (1994), 34.

30. Cited by Martin Gardner, *Science: Good, Bad and Bogus* (Oxford: Oxford University Press, 1983), 163.

31. A good review is given in Robert Buckman and Karl Sabbagh, *Magic or Medicine? An Investigation of Healing and Healers* (London: Macmillan, 1993).

32. Uri Geller, quoted by Marks and Kammann, *The Psychology of the Psychic*, 92.

33. Ibid. 107–9.

34. Psalm 22: 1, 9–10.

35. Fyodor Dostoevsky, *The Idiot*, trans. David Magarshak (Harmondsworth: Penguin, 1971), 446–7.

36. See the final chapter of Humphrey, *Soul Searching*.

Chapter 17. Hello, Aquarius!

First published in the *Guardian*, 16 October 1986.

Chapter 18. Bugs and Beasts Before the Law

Broadcast on BBC Radio 4, November 1986.

1. E. P. Evans, *The Criminal Prosecution and Capital Punishment of Animals* (London: Heinemann, 1906; reprinted, with a foreword by Nicholas Humphrey, London: Faber & Faber, 1987).

2. No one, I gather, has challenged Evans's facts relating to the trials as such, although he has been caught out in one or two minor historical inaccuracies. J. J. Finkelstein, 'The Ox that Gored', *Transactions of the American Philosophical Society*, 71 (1981), Pt. 2, 47–89) concludes that 'a considerable part of the evidence is beyond suspicion'. Nonetheless, Finkelstein among others has raised scholarly objections to Evans's 'style', and in particular to some of his more sweeping historical and psychological pronouncements.

3. Evans, *The Criminal Prosecution and Capital Punishment of Animals*, 1987 reprint, 156.

4. Ibid. 107.

5. Ibid. 114.

6. Ibid. 115.

7. Ibid. 117.

8. Ibid. 118.

9. Ibid. 112.

10. Ibid., Appendix C, 260 (my translation).

11. Ibid. 111.

12. Ibid. 48.

13. Ibid. 154.

14. Ibid. 150.

15. Jean Racine, *The Litigants*, trans. W. R. Dunstan (Oxford: Oxford University Press, 1928).

16. Act III, lines 124 ff.

17. I refer in particular to W. W. Hyde, 'The Prosecution of Animals and Lifeless Things in the Middle Ages and Modern Times', *University of Pennsylvania Law Review*, 64 (1916), 696–730; Finkelstein, 'The Ox that Gored'; Esther Cohen, 'Law, Folklore and Animal Lore', *Past and Present*, 110 (1986), 8–37; see also Jean Vartier, *Les procès d'animaux du moyen âge à nos jours* (Paris: Hachette, 1970).

18. See, for example, Rosalind Hill, *Both Small and Great Beasts* (London: Universities' Federation for Animal Welfare, 1955); Linda Price,

'Punishing Man and Beast', *Police Review*, August 1986; James Serpell, *In the Company of Animals* (Oxford: Blackwell, 1986). Ted Walker's reading of Evans in 1970 inspired an extraordinary poem, republished in *Gloves to the Hangman* (London: Jonathan Cape, 1973).

19. Hyde, 'The Prosecution of Animals and Lifeless Things'.

20. E. Westermarck, *The Origin and Development of Moral Ideas* (London: Macmillan, 1906).

21. Cohen, 'Law, Folklore and Animal Lore'.

22. In relation to supernatural cures for natural plagues, I would quote the great ecologist Charles Elton: 'The affair runs always along a similar course. Voles multiply. Destruction reigns. There is dismay, followed by outcry, and demands to Authority. Authority remembers its experts and appoints some: they ought to know. The experts advise a Cure. The Cure can be almost anything: golden mice, holy water from Mecca, a Government Commission, a culture of bacteria, poison, prayers denunciatory or tactful, a new god, a trap, a Pied Piper. The Cures have only one thing in common: with a little patience they always work. They have never been known entirely to fail. Likewise they have never been known to prevent the next outbreak. For the cycle of abundance and scarcity has a rhythm of its own, and the Cures are applied just when the plague of voles is going to abate through its own loss of momentum.' From C. Elton, *Voles, Mice and Lemmings: Problems in Population Dynamics* (Oxford: Oxford University Press, 1942).

23. 'Indeed, practice provides the strongest argument for connecting the two phenomena, for they exhibit a correlation of time and space. Both animal and witch trials seem to have become increasingly common in Switzerland and the adjoining French and Italian areas during the fifteenth century, and the coincidence is all the more striking because of the almost total absence of any earlier tradition of secular animal trials in Switzerland. Not surprisingly, Switzerland also witnessed the emergence of a hybrid type of process: the trial of an individual animal by a secular court on charges of supernatural behaviour.' Cohen, 'Law, Folklore and Animal Lore'.

24. Robert Darnton, *The Great Cat Massacre* (Harmondsworth: Penguin, 1985), 12.

25. Evans, *The Criminal Prosecution and Capital Punishment of Animals*, 142 and 288.

26. Ibid. 296.

27. Plato, *Laws*, Bk. IX, quoted by Evans, *The Criminal Prosecution and Capital Punishment of Animals*, 173.

28. Ibid.

29. Fyodor Dostoevsky, *The Brothers Karamazov* (1880), Bk. 2, ch. 6.

30. Alexander Pope, *An Essay on Man* (1773), Epistle IV, l. 49.

31. W. B. Yeats, 'The Second Coming' (1921).

32. Sigmund Freud, remarkably enough, seems to have believed the opposite. When a death occurs, he wrote, 'Our habit is to lay stress on the fortuitous causation of the death—accident, disease, infection, advanced age; in this way we betray an effort to reduce death from a necessity to a chance event.'

33. London *Evening Standard*, 1 April 1986.

34. After this chapter was broadcast as a talk in 1986 and Evans's book was reprinted in 1987, there was a flurry of new interest in animal trials. In particular, Julian Barnes lifted a long section from the Appendix to the 1906 edition of the book for his own use in his novel *A History of the World in 10½ Chapters* (London: Cape, 1987), and Leslie Megahey made a feature film about the trial of a pig, *The Hour of the Pig* (BBC Productions, 1994).

35. Cited in Joseph Hemingway, *History of the City of Chester* (Chester: Fletcher, 1831), i. 360–1. 'It is given by Mr. Willett, in his History of Hawarden Parish, who says it is a correct translation from an old Saxon manuscript.'

Chapter 19. Great Expectations

Keynote lecture at the International Congress of Psychology, Stockholm, 2000.

1. Howard Brody, 'The Placebo Response: Recent Research and Implications for Family Medicine', *Journal of Family Practice*, 49 (2000), 649–54; Robert Buckman and Karl Sabbagh, *Magic or Medicine: An Investigation of Healing and Healers* (London: Macmillan, 1993); Christopher C. French, 'The Placebo Effect', in P. Hixenbaugh, R. Roberts, and D. Groome, eds., *Gateways to the Mind: The Psychology of Unusual Experience* (London: Edward Arnold, 2001); Anne Harrington, *The Placebo Effect* (Cambridge, Mass.: Harvard University Press, 1997); Irving Kirsch, *How Expectancies Shape Experience* (Washington, D.C.: American Psychological Association, 1999); J. M. S. Pearce, 'The Placebo Enigma', *Quarterly Journal of Medicine*, 88 (1995), 215–20; Esther M. Sternberg, *The Balance Within: The Science Connecting Health and Emotions* (New York: W. H. Freeman, 2000); Margaret Talbot, 'The Placebo Prescription', *New York Times Magazine*, 9 January 2000; Patrick D. Wall, 'The Placebo and the Placebo Response,' in P. D. Wall and R. Melzack, eds., *Textbook of Pain*, 4th edn. (Edinburgh: Churchill Livingstone, 1999), 1–12.

2. Andrew Weil, *Spontaneous Healing* (London: Warner Books, 1995), 52.

3. Ibid. 53.

4. Buckman, and Sabbagh, *Magic or Medicine*, 246.

5. E. Ernst and K. L. Resch, 'Concept of True and False Placebo Effects', *British Medical Journal*, 311 (1995), 551–3, at 552.

6. Weil, *Spontaneous Healing*, 771 (my italics).

7. Sternberg, *The Balance Within*.

8. G. Zajicek, 'The Placebo Effect Is the Healing Force of Nature', *Cancer Journal*, 8 (1995), 44–5, at 44.

9. Arthur K. Shapiro and Elaine Shapiro, 'The Placebo Effect: Is It Much Ado about Nothing?', in Anne Harrington , ed., *The Placebo Effect* (Cambridge, Mass.: Harvard University Press, 1997), 12–36, at 31.

10. Randolph Nesse and George C. Williams, *Why We Get Sick* (New York: Times Books, 1994).

11. Randolph Nesse, 'What Good Is Feeling Bad?', *The Sciences*, November 1991, 30–7.

12. Ian P. Owens and Ken Wilson, 'Immunocompetence: A Neglected Life History Trait or a Conspicuous Red Herring?', *Trends in Evolution and Ecology*, 14 (1999), 170–2; B. C. Sheldon and S. Verhulst, 'Ecological Immunology: Costly Parasite Defences and Trade-Offs in Evolutionary Ecology', *Trends in Ecology and Evolution*, 11 (1996), 317–21.

13. E. Svensson, L. Råberg, C. Koch, and D. Hasselquist, 'Energetic Stress, Immunosuppression and the Costs of an Antibody Response', *Functional Ecology* (in press).

14. V. A. Olson and I. P. F. Owens, 'Costly Sexual Signals: Are Carotenoids Rare, Risky or Required?', *Trends in Ecology and Evolution*, 13 (1998), 510–14.

15. L. Råberg, M. Grahn, D. Hasselquist, and E. Svensson, 'On the Adaptive Significance of Stress-Induced Immunosuppression', *Proceedings of the Royal Society*, B, 265 (1998), 1637–41.

16. A. P. Møller, P. Christe, and E. Lux, 'Parasitism, Host Immune Function and Sexual Selection', *Quarterly Review of Biology*, 74 (1999), 3–20.

17. Peter Salovey, Alexander J. Rothman, Jerusha B. Detweiler, and Wayne T. Steward, 'Emotional States and Physical Health', *American Psychologist*, 55 (2000), 110–21; and other articles in the special issue on *Happiness, Excellence and Optimal Human Functioning*, ed. Martin E. P. Seligman and Mihaly Csikszentmihalyi, *American Psychologist*, 55 (2000), 1–169.

18. Shlomo Breznitz, 'The Effect of Hope on Pain Tolerance', *Social Research*, 66 (1999), 629–52.

19. Ibid. 647.

20. Nicholas Humphrey, *Soul Searching: Human Nature and Supernatural Belief* (London: Chatto & Windus, 1995); published in the USA as *Leaps of Faith: Science, Miracles, and the Search for Supernatural Consolation* (New York: Basic Books, 1996).

21. Patrick D. Wall, *Pain: The Science of Suffering* (London: Weidenfeld & Nicolson, 1999).

22. Stephen Weinberg, *Dreams of a Final Theory* (London: Vintage, 1994), 207.

Chapter 20. What Shall We Tell the Children?

Oxford Amnesty lecture, 1997. Published in Wes Williams, ed., *The Values of Science: The Oxford Amnesty Lectures 1997* (Oxford: Westview Press, 1998), 58–79, and in *Social Research*, 65 (1998), 777–805.

1. Christopher Cherniak, 'The Riddle of the Universe and its Solution' (1979), reprinted in Douglas R. Hofstadter and Daniel C. Dennett, eds., *The Mind's I* (New York: Basic Books, 1981).

2. Vladimir Mayakovsky, 'I' (1912), in *Mayakovsky and his Poetry*, trans. George Reavey (Bombay: Current Book House, 1955).

3. Statistics from sources quoted in Nicholas Humphrey, *Soul Searching: Human Nature and Supernatural Belief* (London: Chatto & Windus, 1995).

4. National Science Board, *Science and Engineering Indicators, 1996* (Washington, D.C.: US Government Printing Office, 1996).

5. Richard Dawkins, *The Selfish Gene* (Oxford: Oxford University Press, 1976), ch. 11.

6. Jesuit divine (apocryphal).

7. Donald B. Kraybill, *The Riddle of Amish Culture* (Baltimore, Md.: Johns Hopkins University Press, 1989), 119.

8. 'Home Schools: How Do They Affect Children?', *APA Monitor*, December 1996.

9. Court ruling, Iowa, 1985, cited by James G. Dwyer, 'Parents' Religion and Children's Welfare: Debunking the Doctrine of Parents' Rights', *California Law Review*, 82 (1994), 1371–1447, at 1396.

10. John F. Schumaker, *Wings of Illusion* (London: Polity Press, 1990), 33.

11. See, for example, the review by Jerome Kagan, 'Three Pleasing Ideas', *American Psychologist*, 51 (1996), 901–8.

12. Kraybill, *The Riddle of Amish Culture*, 218.

13. Richard Dawkins, 'Viruses of the Mind', in Bo Dahlbom, ed., *Dennett and His Critics* (Oxford: Blackwell, 1993), 13–27.

14. Johan Reinhard, 'Peru's Ice Maidens', *National Geographic*, June 1996, 68–81, at 69.

15. Supreme Court ruling, 1972, cited by Dwyer, 'Parents' Religion and Children's Welfare', 1385.

16. Supreme Court ruling, 1986, cited ibid. 1409.

17. Cited by Carl Sagan, *The Demon-Haunted World* (New York: Random House, 1996), 325.

18. Cited ibid.

19. Supreme Court ruling, 1972, cited by Kraybill, *The Riddle of Amish Culture*, 120.

20. See the extended discussion by Dwyer, 'Parents' Religion and Children's Welfare'; also James G. Dwyer, 'The Children We Abandon: Religious Exemptions to Child Welfare and Education Laws as Denials of Equal Protection to Children of Religious Objectors', *North Carolina Law Review*, 74 (1996), 101–258.

21. Supreme Court ruling, 1972, cited by Kraybill, *The Riddle of Amish Culture*, 120.

22. Tertullian, *The Prescription against Heretics* (2nd century), ch. 7.

23. Adelard of Bath, *Astrolabium* (12th century), quoted in David C. Lindberg, *The Beginnings of Western Science* (Chicago: University of Chicago Press, 1992)

24. Deuteronomy 30: 19.

25. I am indebted for several of the ideas here to James Dwyer, whose critique of the idea of parents' rights stands as a model of philosophical and legal reasoning.

Chapter 21. The Number of the Beast

First published in *New Scientist*, 25 June 1989.

1. Revelation 14: 17–18.

2. Plato, *The Republic*, trans. H. D. P.Lee (Harmondsworth: Penguin, 1955), 315.

Chapter 22. Arms and the Man

Based on an address given at the festival 'A World in Peril', organized by Professions for World Disarmament and held in Southwark Cathedral, London, on 25 October 1986. Published in *New Blackfriar's Journal*, 68 (1987), 35–40.

1. Bertolt Brecht, *Mother Courage* (1938), trans. Eric Bentley (London: Methuen, 1962).

2. Ervin Staub, *The Roots of Evil: The Origins of Genocide and Other Group Violence* (Cambridge: Cambridge University Press, 1989).

3. Matthew 25: 29.

4. Matthew 25: 16–18, 25.

Chapter 23. Death in Tripoli

First published in the *Guardian*, 14 January 1987.

1. William Wordsworth, 'Tax Not the Royal Saint' (1822).

2. Immanuel Kant, *Critique of Judgment* (1790), Bk. I. i. 2, from *Philo-*

sophies of Beauty, ed. and trans. E. F. Carritt (Oxford: Oxford University Press, 1931).

3. Lao-Tzu (6th century BC), trans. Ernest Crosby.

4. William Shakespeare, *Richard II*, V. vi. 39–48.

Chapter 24. *Follow My Leader*

Talk given at the Institute of Historical Research, 6 June 2001, in a seminar on 'Dictatorship: The Seduction of the Masses'.

1. Ian Kershaw, *Hitler, 1936–45: Nemesis* (London: Allen Lane, 2000).

2. Denis Diderot, *Supplement to the Voyage of M. Bougainville* (1772), Pt. IV, in *Diderot, Interpreter of Nature: Selected Writings*, trans. J. Stewart and J. Kemp (London: Lawrence & Wishart, 1937), 187.

3. Plato, *The Republic*, trans. H. D. P. Lee (Harmondsworth: Penguin, 1955), 314.

4. Carl Schmitt, quoted by Andrew Arato, 'Goodbye to Dictatorships?', *Social Research*, 67 (2000), 925–55, at 933.

5. Arato, 'Goodbye to Dictatorships?', 943; compare Heidegger's speech as rector of the University of Freiburg in 1933: 'Not theorems and rules ought to be the rules of your Being. The Führer himself personifies German reality and its law for today and the future.'

6. Joseph Henrich and Robert Boyd, 'Why People Punish Defectors', *Journal of Theoretical Biology*, 208 (2001), 79, and 'The Evolution of Conformist Transmission and the Emergence of Between-Group Differences', *Evolution and Human Behavior*, 19 (1998), 215–41; see also Joseph Henrich and Francisco J. Gil-White, 'The Evolution of Prestige: Freely Conferred Deference as a Mechanism for Enhancing the Benefits of Cultural Transmission', *Evolution and Human Behavior*, 22 (2001), 165–96.

7. John Maynard Keynes, *The End of Laissez-Faire* (London: Hogarth Press, 1926), 3.

Index

Adelard of Bath 315
altruism, evolution of 52–61
Amis, Martin 177
Amish Christians 296–7, 302–7,
    311–12
animal trials 235–54
anthropic principle 230
Aquinas, Thomas 245
Asch, Solomon 336
astrology 217, 232–4
atavisms 192–4
attractors 333
Auden, W. H. 88, 205
Australian Aborigines, exceptional
    memory of 193–4
autism 133, 141, 143–5, 156–8, 191,
    352; *see also* Nadia

Bahn, Paul 151–3, 157–9
Baldwin effect 130, 350
Bartlett, Frederick 190
beauty, masking of 195–6
Becker, Ernst 214
belief, acquisition of 275–7, 281–2
blindsight 69
Bloom, Paul 156, 160
Body of the Dead Christ, The (Holbein)
    228
Borges, Jorge Luis 350
Boyd, Robert 337, 339
Braque, Georges 16–17
Brecht, Bertolt 321
Breznitz, Shlomo 273
Buckman, Robert 259
Byron, Lord 199

CAIS (complete androgen insensitivity
    syndrome) 167
cancer 259–60, 283–4
cave art 132–61
Chalmers, David 67, 93

chaos theory 333–4
Charcot, J. M. 43
charisma 17–18
Chassenée, Bartholomew 237–8, 253
Chaucer, Geoffrey 162–3
Cherniak, Christopher 289
chimpanzees, memory capacity of
    183–7
circumcision, female 293–5, 302,
    309
clocks, evolution of 121–2
Clottes, Jean 134, 141
Cohen, Esther 244–5
consciousness 7–14, 65–85, 90–114,
    115–25
Crick, Francis 100

Damasio, Antonio 100
Darnton, Robert 151–2, 154, 245
Darwin, Charles 65, 70, 76, 130, 165,
    169, 171, 293
da Vinci, Leonardo 17, 162
Dawkins, Richard 53–4, 296, 302,
    314
Deformed Transformed, The (Byron)
    199
democracy 335–6
Dennett, Daniel 19, 67, 98–9, 119,
    145, 155, 159, 314
Derham, William 165–6, 168–70,
    172–3, 197, 199
Descartes, René 65, 98
de Waal, Frans 51
dictatorship 330, 335–9
Diderot, Denis 65–70, 73, 83, 90–1,
    93, 113, 331, 334
dimensions, method of 94–5
Doasyouwouldbedoneby, Mrs 61
dogs, consciousness of 83–4
Donne, John 50–1
Dostoevsky, Fyodor 229, 251

dual currency concepts 99
Dunbar, Robin 146
Dwyer, James 360

Eiffel Tower 163–4
Einstein, Albert 177, 251
Eliot, George 196
Elton, Charles 356
Empson, William 115
Engels, Friedrich 333
Evans, E. P. 235–7, 244, 246, 253
Eysenck, Hans 217

Farrer, Donald 183, 185
fire-making 180–3
flagging the present 117–18
Fodor, Jerry 206
Forster, E. M. 32
Frege, Gottlob 8, 10, 124
Freud, Sigmund 356
Frith, Uta 141, 156, 160

games theory, war and 321–3
Garbo, Greta 18
Geller, Uri 221–4, 226–7

Ghadaffi, death of daughter of 327
Gnostics 50–1
Goethe, J. W. von 93–4
Gombrich, Ernst 132, 158
Grant, Russell 232
green beard effect 54
Greenfield, Susan 100
Gregory, Richard 1, 2, 117
Grew, Nehemiah 172–3, 177–8,
    194–5, 197
Griffin, Donald 76–7

hair-loss 180–3, 190, 350
Hamilton, William 52–3, 55, 59, 61
Hawarden Cross 253–4
Hawking, Stephen 176–7
healing, bodily systems 256–8, 260–3,
    268, 271–2, 281–4
Heidegger, Martin 361
Hemingway, Ernest 208
Henrich, Joseph 337, 339
Hitler, Adolf 330, 336, 339
Holbein, Hans 228–9
Holmes, Sherlock 1
Home, D. D. 222
Hume, David 26
Hyde, W. W. 244

iatrogenesis of MPD 41–2, 44
idiot savants 167, 352; *see also* Nadia
Inca sacrifice 305–6
immune response 260, 267–70, 283–4
impossible triangle 1
inner eye 74–5, 78, 82
intelligence 129–31; limits on 166,
    168, 197; masking of 196–8
inverted spectrum 68, 97, 118
IQ 166, 168, 197; police recruits and
    198

James, William 129
Jesus Christ, psychobiography of
    209–29

Kalahari Bushmen, exceptional mem-
    ory of 193
Kammann, Richard 226
Kant, Immanuel 307, 328
Karamazov, Ivan 251
Kershaw, Ian 330, 339
Keynes, John Maynard 147, 339
Kingsley, Charles 61
kin selection 52–61
Kipling, Rudyard 125
Koch, Christoph 100
Koestler, Arthur 223
Kohut, Heinz 34
Kraybill, Donald 297

language, evolution of 146–51, 154–5,
    189–90; drawing skill and 144–5,
    157; picture memory and 189–90,
    352
Lao-Tzu 328
Lawrence, D. H. 115, 125
learning by doing 324–6
Leibniz, G. W. von 163, 245
Lifton, Robert Jay 34
local maxima, problem of 174
Locke, John 118–19
Luria, Alexander 186–7, 194
Luther, Martin 3

Macphail, Euan 124
Mao Tse Tung 336, 338
Marais, Eugene 28, 193
Marcel, Anthony 117
Marks, David 226
Marlowe, Christopher 87, 162–3
Martin, M. F. 44
Marxism 3, 299, 333, 339

Mary (MPD patient) 22–6, 32–4
match-to-sample task 183–6
Mayakovsky, Vladimir 290
McGinn, Colin 93–4, 97–8
McManus, Christopher 157
memory, masking of 183–94
Milgram, Stanley 337
Miller, Geoffrey 161, 197
mind–body (mind–brain identity) prob-
lem 90–114
Minsky, Marvin 341
Mithen, Steven J. 146–7, 152–3, 156,
161
Mona Lisa 17
Monroe, Marilyn 18
Morris, Desmond 180
mother love 88
Mozart, Wolfgang 162
Multiple Personality Disorder (MPD)
15–18, 19–47

Nadia (autistic girl) 133–60, 191,
193
natural psychologists 77–83
Newson, Elizabeth 150, 160
Newton, Isaac 96, 99, 147, 162–4
Nijinsky 3, 18
Nietzsche, Friedrich 162, 176, 281
number of the beast 318–19

Orwell, George 281
other minds 80–2

pain 103–7, 264–70, 273–4, 277
paranormal, belief in 206–9, 230,
295; *see also* Jesus
Pascal, Blaise 281
Pater, Walter 17
Peirce, C. S. 34
Penrose, Roger 99
Perrott and Warrick Fellowship 206–8
phantasms (qualia) 96
placebo effect 255–85
Plaideurs, Les (Racine) 242–3
Plato 16–18, 29, 48, 249–50, 253,
255, 318–19, 335, 338
Plato's number 318
Pope, Alexander 251
pound auction 322
Povinelli, Daniel 186
privatization of sensation 112–13,
115–25
Proust, Marcel 9

qualia 96–114, 115–25; *see also* sensation

Racine, Jean 242, 249
Ramsey, A. S. 94, 107
Rappoport, Anatol 323
Reagan, Ronald 325
reciprocal altruism 52–61
Reid, Thomas 101, 103, 116–17
Reinhard, Johan 305
religious education 289–317
Repin, Ilya 78–80
Revelation, book of 318–19
rights, children's 289–317
Robert ('mini-Geller') 220–2
Rothstein, Stephen 53–4, 57–8
Russell, Bertrand 206

S. (mnemonist) 186–8, 190,
193–4
Schmitt, Carl 335
Schull, Jonathan 130–1
Schumaker, John 298
scientific education 312–17
self, as actor 12–14, 124; in baby
7–14; as centre of narrative gravity
27–32; in MPD 19–48; in
Shakespeare's friend 16–18
Selfe, Lorna 133, 144–5 152–3
Self-Heal Fairy, The (Barker) 257
sensation, as action 103–7; distin-
guished from perception 100–2,
116–17; evolution of 108 -13,
120–5; functional role of 117–18;
privacy of 112, 118–25
Shah, Amita 141
Shakespeare, William 15–16, 18,
86–8, 162–3, 175, 329
Shapiro, Arthur and Elaine 261
Snow, C. P. 162
Sonnet 53 (Shakespeare) 15
Sonnet 87 (Shakespeare) 8
species mind 130–1
statue, trial of 250
Staub, Ervin 325–6
Steadman, Philip 121
Stockholm syndrome 337
suggestibility 336–9

talents, parable of 326
Tattersall, Ian 143, 153, 183
Taylor, W. S. 44
Tebbit, Norman 327

Tertullian 315
theory of mind 79–83
They Did Not Expect Him (Repin)
  79
thick moment of consciousness 106,
  113
Tolstoy, Leo 196
Trivers, Robert 52–3, 57–9, 61

Valentine, Saint 49–50
Virgin Mary, trial of image of 253
von Neumann, John 321

Wall, Patrick 277

Watson, J. B. 68
Weil, Andrew 258–60
Weinberg, Stephen 281
Wiltshire, Stephen (autistic artist) 153,
  158–9
witness power 204
Wittgenstein, Ludwig 68, 71, 80, 103,
  119
Wordsworth, William 328

Yeats, William Butler 251

Zajicek, George 261
Zubrow, Ezra 152–3